中国Ⅰ级重点保护野生植物手册

Handbook of Wild Flora under Level I Protection in China

张丽荣　潘　哲　孟　锐　金世超　等 编著

中国环境出版集团・北京

图书在版编目（CIP）数据

中国Ⅰ级重点保护野生植物手册/张丽荣等编著. —北京：中国环境出版集团，2022.9

ISBN 978-7-5111-4567-3

Ⅰ. ①中…　Ⅱ. ①张…　Ⅲ. ①珍稀植物—野生植物—中国—手册　Ⅳ. ①Q948.52-64

中国版本图书馆 CIP 数据核字（2020）第 259564 号

出 版 人　武德凯
策划编辑　王素娟
责任编辑　王　菲
封面设计　岳　帅

出版发行　中国环境出版集团
（100062　北京市东城区广渠门内大街 16 号）
网　　址：http://www.cesp.com.cn
电子邮箱：bjgl@cesp.com.cn
联系电话：010-67112765（编辑管理部）
010-67162011（第四分社）
发行热线：010-67125803，010-67113405（传真）
印　　刷　北京中科印刷有限公司
经　　销　各地新华书店
版　　次　2022 年 9 月第 1 版
印　　次　2022 年 9 月第 1 次印刷
开　　本　787×1092　1/16
印　　张　11.75
字　　数　260 千字
定　　价　98.00 元

前言

《国家重点保护野生植物名录（第一批）》（以下简称《名录》）于 1999 年由国务院正式批准公布，这是我国野生植物保护管理工作的一个里程碑。《名录》是由我国野生植物行政主管部门原国家林业局和原农业部共同组织制定的，共列植物 13 类（指种以上分类等级）419 种，其中 Ⅰ 级保护植物有 4 类 67 种，苏铁属和水韭属所有种列为 Ⅰ 级保护。

《中国 Ⅰ 级重点保护野生植物手册》收录了《名录》中的 Ⅰ 级保护植物，由于少数物种的分类学地位存在极大争议或者野外居群一直未被发现，本手册未将异形玉叶金花、密叶红豆杉、元宝山冷杉等物种纳入。本手册共收入种、亚种及变种 77 个。本手册沿用《中国植物志》中科的概念，排列顺序与《名录》相对应，采用《中国植物志》的植物学名。

本手册因撰写时间紧迫，材料收集尚有不足，疏漏与错误难以避免，恳请广大学者和专家批评指正。

作者

2020 年 2 月

目录

1 玉龙蕨/ Yulong Jue

【学　　名】 *Sorolepidium glaciale*

【形态特征】 植株高 10～30 cm，全体密被鳞片或长柔毛，鳞片初为红棕色，老时变为苍白色，卵状或阔披针形，先端纤维状，边缘有睫毛；叶簇生；柄长 4～8 cm，基部直径约 2 mm，褐棕色，向上禾秆色，上面有 2 条纵走沟槽，直通叶轴；叶片线形，长 12～15 cm，宽 2～2.5 cm，一回羽状，羽片约 28 对，互生，近无柄，长圆形，长约 1 cm，宽约 3 mm，圆头，基部对称，近圆形，全缘或略浅裂。叶脉分离，羽状，小脉单一，伸达叶边，通常被鳞毛覆盖，不见。叶厚革质，干后黑褐色，两面密被灰白色的长柔毛，羽轴及主脉下面密被淡棕色，阔披针形，先端纤维状鳞片。孢子囊群圆形，生于小脉顶端，位于主脉与叶边之间，每羽片3～4对，无囊群盖，通常被鳞片所覆盖，不可见。

【孢 子 期】7—10月。

【分　　布】产于四川（稻城、木里）、云南（丽江、香格里拉）、西藏（波密）。中国特有。

【生　　境】生长于海拔3 200～4 700 m高山冰川洞穴、林下岩缝中。

【保护价值】科研。

【致危因素】自然种群过小。

[Scientific Name] *Sorolepidium glaciale*

[Morphological Characters] Plants 10-30 cm tall; scales dense, reddish brown or pale when aged, ovate or broadly lanceolate with fibrous apices and ciliary margin; clustered; stipe 4-8 cm in length, with brown bases 2mm in diam and straw-colored distal with two longitudinal lateral veins directly connected with the rachis; linear lamina 12-15 cm in length and 2-2.5 cm in width, even-pinnate 28 pairs, alternate, attached, sessile, oblong, 1 cm in length and 3 mm in width, apices rounded, bases nearly rounded and symmetrical, margin entire or lobed. Separated veins, pinnate, lateral veins simple extending to the margin, usually covered with scales, invisible. Frond texture thickly leathery, dark brown when dry, densely long gray white cilia on both surfaces, light brown under pinnate rachis and main veins, broadly lanceolate with fibrous scaly apices. Sporangium elliptic, terminal on shorter veinlets，between rachis and margin，even-pinnate 3-4 pairs，ex-indusiate，usually covered with scales，invisible.

[Spore Phase] July to October.

[Distribution] Sichuan Province（Daocheng，Muli），Yunnan Province（Lijiang，Shangri La）and Tibet（Bomê）. A native plant of China.

[Habitat] Frigid alpine mountains，calcareous rock crevices；3,200-4,700 m above sea level.

[Protection Value] Scientific research.

[Threatening Factors] The natural population is too small.

摄影：毛星星

Photo：Mao Xingxing

2 高寒水韭/ Gaohan Shuijiu

【学　　名】*Isoetes hypsophila*

【形态特征】小型蕨类，多年生沼地生植物。植株高不及 5 cm；根茎肉质，块状，长约 4 mm，呈 2～3 瓣裂。叶多汁，草质，线形，长 3～4.5 cm，宽约 1 mm，基部以上鲜绿色，内具 4 个纵行气道围绕中肋，并有横隔膜分隔成多数气室，先端尖，基部广鞘状，膜质，宽约 4 mm。孢子囊单生于叶基部，黄色。大孢子囊矩圆形，长约 3 mm，直径约 2 mm；小孢子囊矩圆形，长约 2.5 mm，直径约 1.5 mm。大孢子球状四面形，表面光滑无纹饰。染色体 $2n=22$。

【孢 子 期】7—10 月。

【分　　布】四川（稻城、九龙）、云南（丽江、宁蒗）。中国特有濒危水生蕨类植物。

【生　　境】生长于海拔约 4 300 m 的高山草甸水浸处。

【致危因素】高原湿地面积逐渐减少和退化，生长环境破坏严重。

[Scientific Name] *Isoetes hypsophila*

[Morphological Characters] Small ferns，perennial marsh plants. Plants less than 5 cm tall; succulent and tuberous rhizome, about 4 mm in length, 2-3 dehiscence. Juiciness, herbaceous and linear leaves, 3-4.5 cm in length and 1 mm in width, vivid green above basal, with 4 longitudinal airways around the midribs, which are partitioned by the cross septa into many air spaces. Acute leaf apices, broad sheath-like basal, membranous, 4 mm in width. Solitary sporangium attached to the leaf basal, elliptic-oblong megasporangium 3 mm in length and 2 mm in diameter; elliptic-oblong microsporangium 2.5 mm in length and 1.5 mm in diameter. Spherical tetrahedron megaspore, smooth walled. Chromosome $2n=22$.

[Spore Phase] July to October.

[Distribution] Sichuan Province (Daocheng, Jiulong), Yunnan Province (Lijiang, Ninglang). Endangered aquatic ferns. A native plant of China.

[Habitat] Alpine meadows at an altitude of about 4,300 m.

[Threatening Factors] The area of plateau wetland is gradually reduced and degraded, and the growth environment is seriously damaged.

摄影：张宪春　周亚东

Photo：Zhang Xianchun　Zhou Yadong

3 东方水韭/ Dongfang Shuijiu

【学　　名】*Isoetes orientalis*

【形态特征】本种近中华水韭，根状茎分裂。叶有 20～40 个簇生，10～20 cm 长，中部宽 2～3 mm，盖膜不完整，仅覆盖孢子囊的上侧边；大孢子表面具网状纹饰，小孢子表面带刺的瘤状突起。大孢子 5—9 月成熟，小孢子 6—10 月成熟。

【孢 子 期】5—10 月。

【分　　布】产于浙江（松阳）。中国特有。

【生　　境】生长于海拔 1 200 m 的沼泽地。

【保护价值】观赏。

【致危因素】生境破碎化和丧失，自然种群过小。

[Scientific Name] *Isoetes orientalis*

[Morphological Characters] Species close to *Isoetes sinensis*，dehiscent rhizome. 20-40 fascicled leaves，10-20 cm in length and 2-3 mm in width at the middle part，incomplete indusium covers only the upper side of the sporangium；megaspore with reticulate ornamentation，microspore with tuberculate protuberance with spiny surface. Megaspores mature from May to September； microspores mature from June to October.

[Spore Phase] May to October.

[Distribution] Zhejiang Province（Songyang）. A native plant of China.

[Habitat] Marshland at an altitude of 1,200 m.

[Protection Value] Ornamental plant.

[Threatening Factors] Fragmented and lost habitats， too small natural populations.

摄影：刘军

Photo：Liu Jun

4 中华水韭/ Zhonghua Shuijiu

【学　　名】*Isoetes sinensis*

【形态特征】多年生沼地生植物，植株高 15～30 cm；根茎肉质，块状，略呈 2～3 瓣，具多数二叉分歧的根；向上丛生多数向轴覆瓦状排列的叶。叶多汁，草质，鲜绿色，线形，长 15～30 cm，宽 1～2 mm，内具 4 个纵行气道围绕中肋，并有横隔膜分隔成多数气室，先端渐尖，基部广鞘状，膜质，黄白色，腹部凹入，上有三角形渐尖的叶舌，凹入处生孢子囊。孢子囊椭圆形，长约 9 mm，直径约 3 mm，具白色膜质盖；大孢子囊常生于外围叶片基的向轴面，内有少数白色粒状的四面形大孢子；小孢子囊生于内部叶片基部的向轴面，内有多数灰色粉末状的两面形小孢子。

【孢 子 期】5—10 月末。

【分　　布】分布于江苏南京，安徽休宁、屯溪和当涂，浙江杭州、诸暨、建德和丽水等地。

【生　　境】主要生长于 10～600 m 的浅水池塘边和山沟淤泥土上。

【保护价值】观赏、科研。

【致危因素】生长环境破坏严重。

[Scientific Name] *Isoetes sinensis*

[Morphological Characters] Perennial marsh plants 15-30 cm in height；succulent and tuberous rhizome，2-3 dehiscence，most roots bifurcational；upward clustering of mostly rachis imbricated leaves. Juiciness，herbaceous and linear leaves，15-30 cm in length and 1-2 mm in width，vivid green above basal，with 4 longitudinal airways around the midribs，which are partitioned by the cross septa into many air spaces. Acute leaf apices，broad sheath-like basal，membranous，yellow-white in color，abdominal indentation，triangular acuminate ligule at the upper end and sporangium at the concave part. Ellipse sporangium，9 mm in length and 3 mm in diameter，with white membranous indusium；megasporangium on the rachis plane of the peripheral leaf basal with a few white granular tetrahedral megaspores；microsporangium on the rachis plane of the inner leaf basal，with most gray two-sided powder microspores.

[Spore Phase] May to late October.

[Distribution] Nanjing in Jiangsu Province，Xiuning，Tunxi and Dangtu in Anhui Province，Hangzhou，Zhuji，Jiande and Lishui in Zhejiang Province.

[Habitat] Shallow water pond and mud soil in the gully，10-600 m.

[Protection Value] Ornamental plant and scientific research.

[Threatening Factors] The growth environment is seriously damaged.

摄影：李晓东　刘军

Photo：Li Xiaodong　Liu Jun

5 台湾水韭/ Taiwan Shuijiu

【学　　名】*Isoetes taiwanensis*

【形态特征】水生至湿生植物。根茎块状，2～4 裂，上部扁平，下部呈圆柱状，基部边缘有薄膜状物质，尖端有气孔散布；叶开展，多汁，草质，鲜绿色，线形，15～90 叶一束，丛生于球茎顶，呈螺旋状排列，长 7～25 cm，具空腔，仅具单脉。叶舌呈三角形延长。孢子囊生长于叶基部内侧。大孢子囊阔椭圆形，表面具皱纹状-网状纹饰。小孢子灰色，椭圆形，具小刺，大孢子湿时呈灰色，干时为白色。

【孢 子 期】7—10 月。

【分　　布】仅见于台湾台北七星山的梦幻湖。中国特有。

【生　　境】生长于海拔 860 m 地带，具干湿双栖性，喜欢生长在浅水地。

【保护价值】观赏、科研。

【致危因素】自然种群过小。

[Scientific Name] *Isoetes taiwanensis*

[Morphological Characters] Aquatic and hygrophyte. Tuberous rhizome , 2-4 dehiscence, flat upper part, cylindrical lower part, with membranous basal edge, pores scattered at the apices; 15-90 fascicled leaves unfolded, juiciness, herbaceous, vivid green and liner, clustered on the top of the corm, arranged in a spiral pattern, 7-25 cm in length, has a cavity, simple vein. Triangular extension of ligule. Sporangium at the inner leaf basal. Broad ellipse megasporangium with wrinkled surface or reticulate ornamentation. Ellipse gray microspore, spinulose; megaspore gray when wet, white when dry.

[Spore Phase] July to October.

[Distribution] Dream Lake in Mount Qixing, Taipei. A native plant of China.

[Habitat] At 860 m above sea level, it has a dry and wet habitat and likes to grow in shallow water.

[Protection Value] Ornamental plant and scientific research.

[Threatening Factors] The natural population is too small.

摄影：张宪春

Photo：Zhang Xianchun

6 云贵水韭/ Yun-gui Shuijiu

【学　　名】 *Isoetes yunguiensis*

【形态特征】 沉水植物，植株高 15～30 cm；根茎短而粗，肉质块状，略呈三瓣，基部有多条白色须根。叶多数，丛生，草质，线形，半透明，绿色，长 20～30 cm，宽 5～10 mm，横切面三角状半圆形，有 4 个薄膜隔和纵行气道，内有长 2～4 mm 的横向隔膜，叶片基部向两侧扩大呈阔膜质鞘状，腹部凹入，其上有三角形叶舌，凹入处生长圆形孢子囊，无膜质盖。植株外围的叶生大孢子囊，大孢子球状四面形，表面具不规则的网状纹饰（网脊不平），直径 360～450 μm。小孢子囊生于内部叶片基部的向轴面，内生多数灰色粉末状小孢子。

【孢 子 期】 7—10 月。

【分　　布】 仅分布于云南昆明、寻甸，贵州平坝。中国特有。

【生　　境】 生长于海拔 1 800～1 900 m 的山沟溪流水中及流水的沼泽地。

【保护价值】 观赏、科研。

【致危因素】 生境被破坏严重。

[Scientific Name] *Isoetes yunguiensis*

[Morphological Characters] Submerged plants 15-30 cm tall；succulent and tuberous rhizome, short and stout, 3 dehiscence, white fibrous roots at the basal. Tufted leaves, herbaceous, liner, translucent and green, 20-30 cm in length and 5-10 mm in width. Triangular semicircle cross section, partitioned by the septa into 4 air spaces, longitudinal airway, cross septa 2-4 mm in length, broad membranous sheath-like leaf basal expands to both sides, abdominal indentation, triangular ligule, round sporangium in the concavity, without membranous, inoperculate. Megasporangia in outer leaves, spherical tetrahedron, irregular reticulate ornamentation on the surface (uneven ridge), 360-450 μm in diameter. Microsporangium on the rachis plane of the inner leaf basal, with gray powdery microspore.

[Spore Phase] July to October.

[Distribution] Kunming and Xundian in Yunnan Province, Pingba in Guizhou Province. A native plant of China.

[Habitat] Mountains, gullies, streams and marshes with flowing water at an altitude of 1,800-1,900 m.

[Protection Value] Ornamental plant and scientific research.

[Threatening Factors] Habitat severely damaged.

摄影：张宪春

Photo：Zhang Xianchun

7 巨柏/ Jubai

【学　　名】*Cupressus gigantea*

【形态特征】乔木，高 30～45 m，胸径 1～3 m，稀达 6 m；树皮纵裂成条状；生鳞叶的枝排列紧密，粗壮，不排成平面，常呈四棱形，稀呈圆柱形，常被蜡粉，末端的鳞叶枝径粗 1～2 mm，不下垂；二年生枝淡紫褐色或灰紫褐色，老枝黑灰色，枝皮裂成鳞状块片。鳞叶斜方形，交叉对生，紧密排成整齐的四列，背部有钝纵脊或拱圆，具条槽。球果矩圆状球形，长 1.6～2 cm，胸径 1.3～1.6 cm；种鳞 6 对，木质，盾形，顶部平，多呈五角形或六角形，或上部种鳞呈四角形，中央有明显而凸起的尖头，能育种鳞具多数种子；种子两侧具窄翅。

【花 果 期】花期 4—5 月，种子翌年夏末秋初成熟。

【分　　布】西藏雅鲁藏布江流域的朗县、米林及林芝等地，甲格以西分布较多。

【生　　境】常见于海拔 3 000～3 400 m 地带，生长于沿江地段的漫滩和有灰石露头的阶地阳坡的中下部。

【保护价值】材质优良，为优良用材树种和产区的造林树种。

【致危因素】大树多被人砍伐，生境遭到破坏。

[Scientific Name] *Cupressus gigantea*

[Morphological Characters] Trees 30-45 m tall, trunk 1-3 m in diameter and rarely 6 m; bark dehiscent longitudinally; branchlets with scale-like leaves densely arranged, stout, non-plane, usually 4-angled, rarely terete, usually glaucous, ultimate ones not drooping, 1-2 mm in diam; biennial branch light purplish brown or grayish purplish brown, old branch dark gray, branch bark split into scales. Scale-like leaves skew square, cross-opposite, closely arranged, in 4 ranks, obtusely ridged or arched gibbous and with a rounded central abaxial gland, 1.6-2 cm in length and 1.3-1.6 cm in diameter; 6 pairs of cone scales, woody, flat apex, often pentagon or hexagon, or quadrilateral upper cone scales, with a prominent, large, free mucro at apex, each fertile scale with numerous seeds, seeds with narrow wings on both sides.

[Florescence and Fruit Period] Florescence April to May, seeds mature in late summer and early autumn of the next year.

[Distribution] Lang County, Milin and Linzhi in the Yarlung Zangbo River Basin, Tibet, more in the west of Jiage.

[Habitat] The middle and lower part of the sunny slope of the flood plain and the terrace with limestone outcrop along the river, with an altitude of 3,000-3,400 m.

[Protection Value] Good material, excellent timber tree and afforestation tree in production area.

[Threatening Factors] Trees are cut down and habitats are destroyed.

摄影：林秦文

Photo: Lin Qinwen

8 宽叶苏铁/ Kuanye Sutie

【学　　名】*Cycas balansae*

【形态特征】树干地下生长，地上部分高 40 cm，直径 35 cm；深褐色，密被鳞片。叶 5～20 片，稀 30 片，一回羽状复叶，长 1.5～3 m，宽 40～60 cm。叶柄在第一年呈绿色，长 20～70 cm，近圆柱形，两侧具 10～25 个刺，刺长 3～8 mm，两边的刺相距 2～6 cm。叶呈阔椭圆形，平直。羽状裂片 20～75 对，每对羽状裂片沿叶轴呈 60°～100°夹角，笔直，长 20～38 cm，宽（1.2）1.8～2.5 cm，纸状，基部收缩成一个很短的小叶柄，边缘直或微呈波浪形，叶尖长而渐尖。鳞片叶（低出叶）三角形，长 4～6 cm，宽 1.2～1.5 cm，褐色且被绒毛。雄球花近圆柱状，长 15～25 cm，宽 4～7 cm；小孢子叶阔楔形，长 1.4～1.7 cm，宽 7～10 mm，密被浅褐色绒毛，中部的小孢子叶先端急尖，稍钝。大孢子叶 5～15（20）片，聚合松散，长 9～13 cm，具浅褐色绒毛，绒毛很少；柄长 5～7 cm；不育顶片阔卵形、近心形或极少呈倒卵形，长 3.5～5.5 cm，宽 2.5～5 cm，深裂为 15～25 个钻状裂片，钻状裂片长 2～3.5 cm，末端裂片较为平整，长 2.5～4 cm；胚珠生于柄的末端，2 或 3 对，光滑无毛。种子通常 2 个，新鲜时淡黄色，干燥后呈棕色宽卵形或椭球形，长 1.8～2.7 cm，径 1.5～2.5 cm；种皮硬质、光滑。授粉期 3—5 月，种子成熟期 9—11 月。

【花 果 期】花期 4—5 月，种期 10—11 月。

【分　　布】生长于广西防城港。越南东北部有分布。

【生　　境】在山谷热带雨林下或石灰石山季雨林中，海拔 100～800 m 的森林中石灰岩衍生土壤的沙地，季雨林红土壤。

【致危因素】盗挖散失，生境破碎化，种群过小。

[Scientific Name] *Cycas balansae*

[Morphological Characters] Trunk subterranean, 40 cm×35 cm above ground , male and female inflorescence; bark dark brown, densely scaly. Leaves 5-20（30）, 1-pinnate, (1.5-3)m×(40-60)cm. Petiole green in 1st year, 20-70 cm, subterete, with 10-25 spines 3-8 mm along each side, 2-6 cm apart, leaf blade oblong, flat. Leaflets in 20-75 pairs, longitudinally inserted at 60°-100°, (20-38)×[(1.2)1.8-2.5]cm, papery, base constricted into a very short petiolule, margin flat or somewhat undulate, apex long acuminate. Scale-like leaves（cataphylls）triangular, (4-6)cm×(1.2-1.5)cm, brown tomentose. Pollen cones subcylindric , (15-25)cm×(4-7)cm ; microsporophylls broadly cuneate , (1.4-1.7)cm×(7-10)mm , pale brown tomentose abaxially , apex of median microsporophylls shortly acute, blunt. Megasporophylls 5-15（20）, loosely grouped, 9～13 cm, pale brown tomentose, glabrescent; stalk 5～7 cm; sterile blade broadly ovate, subcordate, or rarely obovate, (3.5-5.5)cm×(2.5-5)cm, deeply divided into 15-25 subulate, pointed lobes 2-3.5 cm, terminal lobe somewhat flattened, 2.5-4 cm; ovules 2 or 3 on each side of distal part of stalk, glossy, glabrous. Seeds often 2, yellowish when fresh, brown when dry, broadly ovoid or ellipsoid, (1.8-2.7cm)×(1.5-2.5)cm; sclerotesta smooth. Pollination Mar-May, seed maturity Sept-Nov.

[Florescence and Fruit Period] Florescence April to May, seed October to November.

[Distribution] Fangchenggang in Guangxi Zhuang Autonomous Region and Northeast Vietnam.

[Habitat] Under the valley tropical rain forest or in the limestone mountain monsoon forest; deep sand on limestone-derived soil 100-800 m above sea level; lateritic soil in monsoon forests.

[Threatening Factors] Stealing to dig it, fragmented habitat, too small population.

摄影：宋世杰　宋　鼎

Photo：Song Shijie　Song Ding

9 叉叶苏铁/ Chaye Sutie

【学　　名】*Cycas bifida*

【形态特征】树干圆柱形，高 20～60 cm，径 4～5 cm，基部粗 10～12 cm，光滑，暗赤色。叶呈叉状二回羽状深裂，长 2～3 m，叶柄两侧具宽短的尖刺；羽片间距离约 4 cm，叉状分裂；裂片条状披针形，边缘波状，长 20～30 cm，宽 2～2.5 cm，幼时被白粉，后呈深绿色，有光泽，先端钝尖，基部不对称。雄球花圆柱形，长 15～18 cm，径约 4 cm，梗长 3 cm，粗 1.5 cm；小孢子叶近匙形或宽楔形，光滑，黄色，边缘橘黄色，长 1～1.8 cm，宽约 8 mm，顶部不育部分长约 8 mm，有绒毛，圆或有短而渐尖的尖头，花药 3～4 个聚生；大孢子叶基部柄状，橘黄色，长约 8 cm，柄与上部的顶片近等长或稍短，胚株 1～4 枚，着生于大孢子叶柄的上部两侧，近圆球形，被绒毛，上部的顶片菱状倒卵形，宽约 3.5 cm，边缘具篦齿状裂片，裂片钻形，直立，长 1.5～2 cm。种子成熟后变黄，长约 2.5 cm。

【花 果 期】花期 4—5 月，种期 9—10 月。

【分　　布】产于广西龙津、云南。越南北部也有分布。

【生　　境】生长于海拔 100～200 m 地带，常绿、落叶混交林、竹林中或石山灌丛半阴处。

【保护价值】观赏、科研。

【致危因素】生境因农垦破碎化、盗挖散失严重。

[Scientific Name] *Cycas bifida*

[Morphological Characters] Trunk cylindrical to 20-60 cm above ground，4-5 cm in diameter，basal 10-12 cm in diameter，smooth dark red. 2-pinnate forked leaves，deeply divided，2-3 m in length，both sides of petiole with broad short spines；gap between pinnae is about 4 cm，forked split，strip leaves lanceolate，margin undulate，(20-30)cm×(2-2.5)cm，glaucous when young，then dark green，gloss，apex blunt ， base asymmetrical. Pollen cones subcylindric，15-18 cm in length，4 cm in diameter，stalk 3 cm×1.5 cm ； microsporophylls broadly cuneate or spoon shaped，smooth，yellow，orange margin，(1-1.8)cm×8 mm，sterile blade 8 mm，tomentose，apex round or short acuminate，3-4 clustered anthers；megasporophylls base constricted into a petiolule，orange，8 cm in length，nearly equal to or slightly shorter than the terminal lobes，ovules 1-4 on each side of distal part of stalk，spherical，tomentose，terminal lobes rhomboid obovate，3.5 cm in width，subulate lobes on the margin，straight，1.5-2 cm. Seeds turn yellow when mature，2.5 cm.

[Florescence and Fruit Period] Florescence April to May，seed September to October.

[Distribution] Longjin in Guangxi Zhuang Autonomous Region，and Yunnan Province. Also found in North Vietnam.

[Habitat] Evergreen and deciduous mixed forest，bamboo forest or half shade of rocky mountain shrub with altitude of 100-200 m.

[Protection Value] Ornamental plant and scientific research.

[Threatening Factors] Habitat is seriously fragmented and lost due to farming and stealing.

摄像：林秦文

Photo：Lin Qinwen

10 葫芦苏铁/ Hulu Sutie

【学　　名】*Cycas changjiangensis*

【形态特征】茎干常地下生，圆柱状或葫芦状，有时呈串珠状，基部常骤然膨大，径约 25 cm，地上茎高达 50 cm，径 15 cm，顶端近无毛；干皮灰色，下部近光滑。叶 25～45 片，一回羽裂，长 50～130 cm；叶柄横断面近圆形，长 10～40 cm，具刺 9～16 对；羽片与叶轴夹角 55°～60°，革质，长 10～17（23）cm，宽 4～7（9）mm，基部下延，边缘平或微反曲，中脉上面明显隆起，下面干时近平。小孢子叶球圆锥状圆柱形，长 15～25 cm，直径 4～6 cm；小孢子叶楔形或宽楔形，长 1.5～2 cm，宽 0.6～1 cm，先端近全缘，中央具短突尖头。大孢子叶长 8～13 cm，密被黄褐色绢质绒毛；不育顶片宽卵形或扇形，长 5～6 cm，宽 4～8 cm，裂片钻状，每侧 8～17 个，长 2～3.8 cm，顶生裂片宽披针形，长 1.5～3.5 cm，边缘具 1～2 枚齿；胚珠 2～4 枚，无毛。种子 2～4 枚，绿色，熟时变黄褐色，宽倒卵状或近球状，长约 2 cm。孢子叶球期 4—5 月，种子 10—11 月成熟。

【花 果 期】花期 3—4 月，种期 9—10 月。

【分　　布】分布于海南昌江。中国特有。

【生　　境】生长于海拔 800～900 m 季节性干燥旱杂木林或灌丛沙壤土上或火烧迹地。

【保护价值】观赏。

【致危因素】生境退化和丧失；直接采挖；物种内在因素：生境退化和丧失。

[Scientific Name] *Cycas changjiangensis*

[Morphological Characters] Trunk often subterranean, cylindric or flask-shaped, sometimes moniliform, to 50 cm×15 cm, base abruptly swollen to 25 cm in diam, apex almost glabrous; bark gray, nearly smooth toward base of trunk. Leaves 25-45, 1-pinnate, 50-130 cm in length; petiole suborbicular in cross section, 10-40 cm, with 9-16 spines along each side throughout length; leaf blade longitudinally inserted at 55°-65° to rachis10-17 (23) cm×4-7 (9) mm, leathery, base decurrent, margin flat or slightly recurved, midvein prominent on both surfaces, base flat. Pollen cones conical-cylindric, (15-25)cm×(4-6)cm; microsporophylls cuneate or broadly cuneate, (1.5-2)cm×(0.6-1)cm, apex subentire, with short, upcurved mucro. Megasporophylls 8-13 cm, densely yellowish brown silky tomentose; sterile blade broadly ovate or flabellate, (5-6)cm×(4-8)cm, margin pectinate, with 8-17 lobes on each side, 2-3.8 cm long, terminal lobe broadly lanceolate, 1.5-3.5 cm, margin irregularly serrulate or biserrulate, with 1-2 dentate lobes; ovules 2 or 4 on each side of distal part of stalk, glabrous; seeds 2-4, green or yellowish brown when mature, broadly obovoid or subglobose, 2 cm. Strobile Apr-May, seed maturity Oct-Nov.

[Florescence and Fruit Period] Florescence March to April, seed September to October.

[Distribution] Changjiang in Hainan Province. A native plant of China.

[Habitat] Seasonal dry mixed forest or shrub on sandy loam or burned land, 800-900 m above sea level.

[Protection Value] Ornamental plant.

[Threatening Factors] Habitat degradation and loss; direct digging; intrinsic factors of species: habitat degradation and loss.

摄像：张敬莉　周洪义

Photo: Zhang Jingli　Zhou Hongyi

11 德保苏铁/ Debao Sutie

【学　　名】*Cycas debaoensis*

【形态特征】德保苏铁叶为三回羽裂、茎干地下生等与多歧苏铁相似，其不同主要在于：叶（3）5～11（15）片，长 1.5～2.7 m，叶柄长 0.6～1.3 m；羽片线形（初生的第一片叶为倒卵状披针形，先端常尾状渐尖），长 10～22（28）cm，宽 0.8～1.5 cm，先端渐窄或长渐尖；小孢子叶窄楔形，长 3～3.5 cm；大孢子叶长 15～20 cm，不育顶片绿色，近心形或近扇形，每侧具裂片 19～25 个，丝状，长 3～6 cm；胚珠 4～6 枚；种子 3～4 个，倒卵状球形，长 3～3.5 cm。孢子叶球期为 3—4 月，种子 11 月成熟。

【花 果 期】花期 4—5 月，种期 10—11 月。

【分　　布】产于广西德保、云南富宁。中国特有。

【生　　境】生长于海拔 600～1 100 m 的石灰岩山地向阳矮灌丛和海拔 200～600 m 的砂岩、页岩常绿阔叶林下。

【保护价值】观赏、科研。

【致危因素】直接采挖；物种内在因素；生境退化和丧失。

[Scientific Name] *Cycas debaoensis*

[Morphological Characters] Leaves 3-pinnate，trunk often subterranean like Cycas multipinnata. The main difference is：Leaves（3）5-11（15），1.5-2.7 m in length，petiole 0.6-1.3 m；liner leaflets（primary leaflets obovate lanceolate，apex acuminate），(10-22)（28）cm× (0.8-1.5)cm，apex long attenuate or long acuminate；microsporophylls narrowly cuneate，3-3.5 cm；megasporophylls 15-20 cm，sterile blade green，subcordate or subflabellate，19-25 filiform lobes 3-6 cm；ovules 4 to 6；seeds 3 to 4，subglobose or obovoid-globose，3-3.5 cm in length. Strobile Mar-Apr，seed maturity Nov.

[Florescence and Fruit Period] Florescence April to May，seed October to November.

[Distribution] Debao in Guangxi Zhuang Autonomous Region，Funing in Yunnan Province. A native plant of China.

[Habitat] Thickets and sparse forests of small trees on dry，sunny，open slopes of limestone hills，600-1,100 m above sea level；evergreen broad-leaved forest on sandstone or shale，200-600 m above sea level.

[Protection Value] Ornamental plant and scientific research.

[Threatening Factors] Direct digging；intrinsic factors of species：habitat degradation and loss.

摄像：宋　鼎　华国军

Photo：Song Ding　Hua Guojun

12 滇南苏铁/ Diannan Sutie

【学　　名】*Cycas diannanensis*

【形态特征】滇南苏铁树干高 0.8～3 m，直径 30～40 cm，具环状叶痕，下部渐脱落。羽叶长 2.5～3 m，羽片 67～138 对，纸质叶柄长 0.8～1 m，两侧具短刺 36～40 对，先端尖锐而微弯，长 0.2～0.4 cm，羽片条形，中部羽片长 25～38 cm，宽 1.4～1.5 cm，中脉两面隆起。雄球花柱状卵圆形，小孢子叶长约 4 cm，上面不育部分被黄褐色绒毛。雌球花卵圆形，长约 20 cm，直径 15 cm；大孢子叶长 26～80 cm，顶片宽圆形或卵圆形，背面密被黄褐绒毛，腹面无毛，边缘具 13～20 对钻形裂片，长 2～3.5 cm，大孢子叶柄长（3）6～17.5 cm，两侧着生 2～7 枚胚珠，胚珠疏生毛与无毛并存。种子球形，熟时红褐色，长 3～6 cm，直径 2.1～2.8 cm。

【花 果 期】花期 4—5 月，种期 10 —11 月。

【分　　布】分布于云南省个旧市南部绿水河与清水河之间蔓耗镇附近。中国特有。

【生　　境】海拔 700～1 200 m 陡峭的石灰岩山地草丛或阔叶林下，呈星散或小片状分布。

【保护价值】观赏。

【致危因素】生境破碎化，盗伐。

【附　　注】缺少影像资料。

[Scientific Name] *Cycas diannanensis*

[Morphological Characters] Trunk 0.8-3 m tall，30-40 cm in diameter，with leaf scars，subsmooth toward base. Leaves 2.5-3 m，leaflets in 67-138 pairs，papery petiole 0.8-1 m，with 36-40 pairs of short spines along each side throughout length，apex acuminate and slightly curved，0.2-0.4 cm in length，leaflets liner，middle leaflets (25-38)cm×(1.4-1.5)cm，midvein raised on both surfaces. Pollen cones ovoid，microsporophylls 4 cm，apical sterile part yellow-brown tomentose. Pistil ellipsoid，20 cm in length and 15 cm in diameter;

megasporophylls 26-80 cm, terminal lobes broadly rounded or ellipsoid, yellow-brown tomentose abaxially, glabrous adaxially, divided into 13-20 subulate at the margin, 2-3.5 cm, petiole (3) 6-17.5 cm, with ovules 2 or 7 on each side of it, glabrous or thinly covered tomentose. Seeds globose, reddish-brown when mature, 3-6 cm in length, 2.1-2.8 cm in diameter.

[Florescence and Fruit Period] Florescence April to May, seed October to November.

[Distribution] Near Manhao Town, between Lyushui River and Qingshui River, South Gejiu City, Yunnan Province. A native plant of China.

[Habitat] Distributed in scattered or small pieces under the grass or broad-leaved forest in the limestone mountain area with an altitude of 700-1,200 m.

[Protection Value] Ornamental plant.

[Threatening Factors] Habitat fragmentation, illegal logging.

[Note] No image data.

13 长叶苏铁/ Changye Sutie

【学　　名】*Cycas dolichophylla*

【形态特征】茎干高 0.8～1.5 m，8～40 枚羽叶集生干顶。羽叶长 2～3.5 m，两侧缘稍下弯；羽片被红褐色的绒毛，短而宽；叶柄光滑，长 40～100 cm，几乎全柄有刺，刺长 4～6 mm；中部羽片长 28～35 cm，宽 1.4～2.2 cm，基部圆，边缘波状，叶脉上凸下平。鳞叶三角状披针形，长 8～12 cm，柔软，被细毛。小孢子叶球窄长卵圆状或纺锤状，长 35～50 cm；小孢子叶柔软，长 3～5 cm，背腹面不加厚，顶端钝圆。大孢子叶不育顶片圆形，长 6～12 cm，被褐色绒毛，边缘具 16～26 对裂刺，顶裂刺与侧裂刺相似。种子卵状，长 3.8～4.2 cm，黄色，中种皮疣状。

【花 果 期】花期 4—5 月，种期 10—11 月。

【分　　布】分布于广西德保、云南。越南北部有分布。

【生　　境】常见于低海拔常绿阔叶林或灌丛中。

【保护价值】观赏。

【致危因素】生境退化和丧失；直接采挖；物种内在因素：生境退化和丧失。

【附　　注】缺少影像资料。

[Scientific Name] *Cycas dolichophylla*

[Morphological Characters] Stem 0.8-1.5 m，with 8-40 leaflets at the top in clusters. Leaflets 2-3.5 m，lateral margin slightly downward curved；reddish-brown tomentose，short and broad；petiole smooth，40-100 cm，with spines 4-6 mm all over；middle leaflets (28-35)cm×(1.4-2.2)cm，base rounded，margin undulate，vein upper convex lower flat. Cataphylls triangular，lanceolate，8-12 cm，soft，tomentose. Microsporophylls ovoid or fusiform，35-50 cm in length；microsporophylls soft，3-5 cm，non-thickened，apical part broadly rounded. Megasporophylls apical sterile part rounded，6-12 cm in length，brown tomentose，margin divided into 16-26 filiform lobes，terminal lobes similar to side lobes.

Seeds ellipsoid，3.8-4.2 cm in length，yellow，sclerotesta verrucose.

[Florescence and Fruit Period] Florescence April to May，seed October to November.

[Distribution] Debao in Guangxi Zhuang Autonomous Region，Yunnan Province. Northern Vietnam.

[Habitat] Evergreen broad leaved forest or shrub at low altitude.

[Protection Value] Ornamental plant.

[Threatening Factors] Habitat degradation and loss；direct digging；intrinsic factors of species；habitat degradation and loss.

[Note] No image data.

14 仙湖苏铁/ Xianhu Sutie

【学　　名】*Cycas fairylakea*

【形态特征】仙湖苏铁树干圆柱形，高可达 1～1.5 m，直径 20～30 cm，有的树干不明显，叶痕宿存；鳞叶披针形，长 8～13 cm，宽 1.5～2.5 cm；羽叶多数，长 2～3.1 m，叶柄长 0.6～1.3 m，光滑或被污毛，具刺 29～73 对，刺长 0.2～0.5 cm，间距 1～4 cm，幼叶锈色；羽片 66～113 对，中部羽片长 17～39 cm，宽 0.8～1.7 cm，间距 2～2.5 cm，羽片平展，边缘平至微反卷，有时波状，条形至镰刀状，薄革质至革质，上面深绿色，有光泽，下面浅绿色，中脉两面隆起；小孢子叶球圆柱状长椭圆形，长 35～60 cm，径 5.5～10 cm，小孢子叶楔形，长 1.8～3 cm，不育部分菱状椭圆形，密被褐色短绒毛，长 0.6 cm，宽 1.2～1.8 cm，顶端具 0.2～0.3 cm 的小尖头，每侧常具 1～4 个小齿，长达 0.1 cm；大孢子叶球半球形，径 35 cm，高 15 cm，大孢子叶长 10～19 cm，密被黄褐色绒毛，后逐渐脱落，仅柄部有残留，顶片卵圆形至卵状披针形，长 5～8.5 cm，宽 5～8 cm，边缘篦齿状深裂，侧裂片 13～24 枚，长（2）2.3～3.6 cm，顶裂片钻形至披针形，明显长于侧裂片，胚珠（2）4～6（8）枚，扁球形，无毛，直径 0.5～0.7 cm，先端有短尖头；种子倒卵状球形至扁球形，黄褐色，长 3～6 cm，直径 2.6～3 cm，无毛，中种皮具疣状突起。

【花 果 期】花期 4—5 月，种期 8—10 月。

【分　　布】产于广东省。中国特有。

【生　　境】生长于海拔 100～500 m 的山地草丛或阔叶林下，呈星散或小片状分布。

【保护价值】观赏。

【致危因素】生境退化和丧失；直接采挖；物种内在因素；生境退化和丧失。

【附　　注】缺少影像资料。

[Scientific Name] *Cycas fairylakea*

[Morphological Characters] Trunk cylindrical, 1-1.5 m tall, 20-30 cm in diameter, some not obvious, with leaf scars; scale-like leaves lanceolate, (8-13) cm×(1.5-2.5) cm; leaflets 2-3.1 m, petiole 0.6-1.3 m, smooth or tomentose, spines 29-73 pairs, 0.2-0.5 cm in length, 1-4 cm apart, young leaves rust; leaflets 66-113 pairs, middle leaflets (17-39) cm×(0.8-1.7) cm, 2-2.5 cm apart, leaflets blade flat, margin flat to slightly reverse curved, somewhat undulate, liner to sickle shape liner, slightly leathery or leathery, upper dark green, gloss, lower light green, midvein raised on both surfaces; Pollen cones cylindrical long ellipsoid, 35-60 cm, 5.5-10 cm in diameter, microsporophylls cuneate, 1.8-3 cm, apical sterile part subrhombic, densely brown tomentose, 0.6×(1.2-1.8) cm, with acute apex 0.2-0.3 cm, with 1-4 lobes 0.1 cm in length on both sides, megastrobile hemispheric, 35 cm in diameter, 15 cm tall, megasporophylls 10-19 cm, densely yellowish-brown tomentose, gradually smooth when mature with remaining tomentose petiole, terminal lobes ellipsoid to ellipsoid lanceolate, (5-8.5)cm×(5-8)cm, deeply divided into 13-24 side lopes at the margin, (2) 2.3-3.6 cm, terminal lobe subulate to lanceolate, longer than side lobes, ovules (2) 4-6 (8), oblate spheroid, glabrous, 0.5-0.7 cm in diameter, apex short acuminate; seeds obovoid or subglobose, yellowish-brown, 3-6 cm in length, 2.6-3 cm in diameter, glabrous, sclerotesta verrucose.

[Florescence and Fruit Period] Florescence April to May, seed August to October.

[Distribution] Guangdong Province. A native plant of China.

[Habitat] Under the mountain grass or broad-leaved forest with an altitude of 100-500 m, distributed in scattered or small pieces.

[Protection Value] Ornamental plant.

[Threatening Factors] Habitat degradation and loss; direct digging; intrinsic factors of species; habitat degradation and loss.

[Note] No image data.

15 锈毛苏铁/ Xiumao Sutie

【学　　名】*Cycas ferruginea*

【形态特征】茎干近一半地下生，地上茎高达 60 cm，直径 15 cm，基部常膨大，中部常有数个稍凹陷的环；干皮灰色，上部有宿存叶基，下部近光滑。叶 25～40 片，回羽裂，长 1～2 m，宽 40～60 cm；叶柄长 45～70 cm，具刺（0）8～21 对，刺毛约 3 mm，叶轴（尤一年生的背面）密被锈色绒毛；羽片与叶轴间夹角近 90°，叶间距 1～2 cm，直或下弯，薄革质，长 20～28 cm，宽 6～10 mm，基部渐窄，具短柄，对称，几不下延，边缘强烈反卷，下面被锈色绒毛，中脉上面明显隆起，干时中央常有 1 条不明显的槽，下面稍隆起。小孢子叶球卵状纺锤形，长 20～35 cm，径 6～10 cm；小孢子叶宽楔形，长 1.5～3 cm，宽 1.2～1.5 cm，顶端边缘有少数浅齿，先端具上弯的短尖头。大孢子叶长 9～14 cm，不育顶片菱状卵形，长 3.5～5.5 cm，宽 3～4.5 cm，裂片每侧 8～15 枚，钻形，长 1～2.5 cm，顶生裂片 3～4 cm；胚珠 4～6 枚，无毛。种子 2～4 个，黄至橘红色，倒卵状球形或近球形，长 2～2.8 cm，中种皮光滑。孢子叶球期在 3—4 月，种子在 9—10 月成熟。

【花 果 期】花期 3—4 月，种期 8—10 月。

【分　　布】产于广西田东。越南北部也有分布。

【生　　境】生长于海拔 200～500 m 山坡灌丛石灰岩峭壁上。

【保护价值】观赏。

【致危因素】盗挖散失，生境破碎化，种群过小。

[Scientific Name] *Cycas ferruginea*

[Morphological Characters]Trunk half subterranean, 60 cm×15 cm above ground, base swollen, often several slightly concave rings in the middle; bark gray, scaly toward apex of trunk, nearly smooth toward base. Leaves 25-40, 1-pinnate, (1-2)m×(40-60)cm; petiole 45-70 cm, with (0) 8-21 spines 3 mm along each side, densely rusty brown tomentose (especially abaxially annual leaves); leaflets longitudinally inserted at 90° to rachis, 1-2 cm apart, straight or downcurved, subleathery, (20-28) cm×(6-10) mm, base attenuate, symmetric, straight, margin strongly recurved, with rusty brown tomentose, midvein raised on both surfaces, usually an inconspicuous groove in the center when dry, raised slightly abaxially, microstrobile ovoid spindle, 20-35 cm long, 6-10 cm in diameter. Microsporophylls broadly cuneate, (1.5-3) cm×(1.2-1.5) cm, a few shallow dentate lobes at the top margin, apex with short mucro upwards curved slightly. Megasporophylls 9-14 cm in length, apical sterile part rhombus ovate, (3.5-5.5) cm×(3-4.5) cm, divided into 8-15 subulate at each margin, 1-2.5 cm in length, terminal lobe 3-4 cm; ovules 4-6, glabrous. Seeds 2-4, yellow to orange, subglobose or obovoid-globose, 2-2.8 cm, sclerotesta smooth. Pollination Mar-Apr, seed maturity Sept-Oct.

[Florescence and Fruit Period] Florescence March to April, seed August to October.

[Distribution] Tiandong in Guangxi Zhuang Autonomous Region. Also found in Northern Vietnam.

[Habitat] A bush or limestone cliff on a hillside 200-500 m above sea level.

[Protection Value] Ornamental plant.

[Threatening Factors] Stealing, habitat fragmentation, population is too small.

摄像：李西贝阳　张敬莉

Photo: Li Xibeiyang　Zhang Jingli

16 贵州苏铁/ Guizhou Sutie

【学　　名】*Cycas guizhouensis*

【形态特征】树高近 1 m，具大型羽状叶，叶柄长 40～50 cm，两侧具短刺，刺长近 3 mm；羽状裂片条形或条状披针形，厚革质，长 8～18（29）cm，宽 8～12 mm，无毛，基部两侧稍有不对称，先端渐尖，边缘稍反曲，表面深绿色，略有光泽，背面淡绿色，中脉在两面隆起。大孢子叶在茎顶密生呈球形，密被黄褐色绒毛或锈褐色绒毛，长 14～20 cm，顶片近圆形，深羽裂，长 6～7 cm，宽 7～8 cm；钻形裂片 17～33 枚，长 2～4.5 cm，宽 2～4 cm，先端渐尖，两面无毛，边缘和基部密生黄褐色绒毛，顶和裂片长 3～4.5 cm，宽 1～1.7 cm，下部有 3～5 线裂片；大孢子叶的下部有急缩成粗短的柄，长 3～5 cm，两侧着生胚珠 2～8 枚；胚珠无毛，球形或近球形，稍扁，肉黄色，顶端红褐色，具短的小尖头。

【花 果 期】花期 4—6 月，种期 9—10 月。

【分　　布】分布于广西隆林，贵州安龙、兴义，云南。中国特有。

【生　　境】生长于海拔 400～1 000 m 石灰岩山坡混交林或灌丛中。

【保护价值】观赏。

【致危因素】盗挖散失严重，农垦使生境破碎化，种群过小。

【附　　注】本种曾被错误地归并到四川苏铁（*Cycas szechuanensis*），形态学与分子生态学证据支持其恢复为独立的种。

[Scientific Name] *Cycas guizhouensis*

[Morphological Characters] Trunk 1 m，large pinnate leaves，petiole 40-50 cm，with short spines 3 mm along each side；pinnate lobes liner or lanceolate，thick leathery，[8-18(29)]cm×(8-12) mm，glabrous，base slightly asymmetrical，apex acuminate，margin slightly recurved，surface dark green，slightly gloss，back light green，midvein raised on both surfaces. Megasporophylls tightly grouped at stalk top，spherical，yellowish brown or rust brown tomentose，14-20 cm in length，terminal lobe suborbicular，with pinnate dehiscence (6-7) cm×(7-8) cm，deeply divided into 17-33 subulate lobes (2-4.5) cm×(2-4) cm，apex acuminate，glabrous on both surfaces，dense yellowish brown tomentose at margin and base，terminal lobe (3-4.5) cm×(1-1.7) cm，with 3-5 liner lobes at the lower part；base constricted into a short petiolule under megasporophylls，3-5 cm，ovules 2 or 8 on each side；ovules glabrous，subglobose or globose，slightly flattened，yellow，upper part reddish brown，with short apex.

[Florescence and Fruit Period] Florescence April to June，seed September to October.

[Distribution] Longlin in Guangxi Zhuang Autonomous Region，Xingyi and Anlong in Guizhou Province，Yunnan Province. A native plant of China.

[Habitat] Mixed forest or shrub of limestone hillside with an altitude of 400-1,000 m.

[Protection Value] Ornamental plant.

[Threatening Factors] Serious theft loss，fragmentation of habitat and small population due to farming.

[Note] This species has been wrongly classified as *Cycas szechuanensis*，while morphological and molecular ecological evidence support its restoration to an independent species.

摄像：林秦文

Photo：Lin Qinwen

17 海南苏铁/ Hainan Sutie

【学　　名】*Cycas hainanensis*

【形态特征】羽状叶长约 1 m，叶柄长约 20 cm、横切面四方状圆形，两侧密生刺，刺间距离约 1 cm，刺长 3～4 mm；羽状裂片近对生，条形，革质，斜上伸展，中部的羽状裂片与叶轴约成 60°，通常直或微弯，长约 15 cm，宽约 6 mm，先端渐尖，基部收缩不对称，下侧下延生长，边缘微向下反卷，上面深绿色，有光泽，下面淡绿色，两面光滑，上面的中脉显著隆起，下面的中脉微隆起。大孢子叶幼时呈褐色绒毛，后渐脱落几无毛，上部的顶片斜方状卵形，长约 7 cm，宽约 5 cm，边缘羽状分裂，每边有裂片 5～7 条，裂片条状钻形，长 2～3 cm，粗约 2.5 mm，先端有刺状尖头，顶生裂片矩圆形，扁平，上部变坚硬，长 3.5～4 cm，宽 1.5～2 cm，边缘在中下部全缘，在上部具数枚锯齿或再分裂，先端有一长刺，长 1～1.4 cm，边缘有少数不明显的疏锯齿；大孢子叶的下部柄状，长约 7 cm，横切面四方状圆形，在其中部两侧着生 2 枚胚珠；胚珠卵圆形，无毛。

【花 果 期】花期 3—5 月，种期 9—10 月。

【分　　布】生长于海南。中国特有。

【生　　境】常生长于海拔 1 200 m 以下的热带或亚热带雨林中。

【保护价值】观赏。

【致危因素】农垦、生境破碎化。

【附　　注】缺少影像资料。

[Scientific Name] *Cycas hainanensis*

[Morphological Characters] Pinnate leaves 1 m，petiole 20 cm，square and round cross section，with dense spines 3-4 mm along each side，1 cm apart；pinnate lobes subopposite，liner，leathery，obliquely upward，median lobes inserted at 60° to rachis，straight or slightly curved，15 cm×6 mm，apex acuminate，asymmetric base constricted，decurrent，margin slightly recurved，adaxial surface dark green and glossy，abaxially light green，both surfaces smooth, midvein strongly raised on adaxial surface, slightly raised on back surface. Megasporophylls brown tomentose when young, glabrescent, terminal lobe square ovoid, 7 cm×5 cm, pinnate lobes at margin, with 5-7 liner lobes on each side, subulate lobes (2-3) cm×2.5 mm, spinous apex, elliptic-oblong terminal lobe, flat, turns stiff at the upper part, (3.5-4) cm×(1.5-2) cm, margin complete at the middle and lower part, and divided into lobes at the upper part, apex long spine 1-1.4 cm, with inconspicuous sparse serration. Megasporophylls petiolule 7 cm, square and round cross section, ovules 2 on each side of middle part of stalk; ovules ellipsoid, glabrous.

[Florescence and Fruit Period] Florescence March to May, seed September to October.

[Distribution] Hainan Province. A native plant of China.

[Habitat] Tropical or subtropical rain forest below 1,200 m above sea level.

[Protection Value] Ornamental plant.

[Threatening Factors] Land reclamation and habitat fragmentation.

[Note] No image data.

18 灰干苏铁/ Huigan Sutie

【学　　名】*Cycas hongheensis*

【形态特征】灰干苏铁树干高 5～7 m，直立，有时分枝，基部膨大，主干直径 30～40 cm，有时可达 1.3 m，树皮纵裂，光滑，灰白色，有时具凸出的同心环，无干顶被毛，叶基仅在干预宿存。鳞叶披针形，坚硬，长 3～4.5 cm，基部宽 1～1.5 cm，顶端刺尖，上面光滑，背面密被黄褐色绒毛。羽叶长 50～90（120）cm，V 字形开展，羽片 49～58 对，对生或近对生，幼时被黄褐色柔毛，后逐渐脱落，叶柄长 10～26 cm，中部以上具 6～22 对短刺，刺长 0.1～0.2 cm，刺距 0.5～3.5 cm，中部羽片长 7～16 cm，宽 0.6～0.8 cm，边缘平或微反卷，革质，上面淡绿色，有光泽，下面黄绿色，中脉上面平，下面隆起；羽片微翘成龙骨状，叶柄、叶轴被黄褐色短柔毛；小孢子叶球圆柱形，长 18 cm，直径 4.5 cm，黄褐色，小孢子叶楔形，长 1.5～1.8 cm，不育部分盾状，宽 0.8～1.1 cm，密被短柔毛，先端有小尖头；大孢子叶顶片卵形至椭圆状披针形，长 4.5～8.5 cm，宽 3～5.5 cm，两侧各具（6）10～16 对深裂片，裂片钻形，纤细，中部裂片长 1.5～2.5 cm，基部粗 1～2 cm，有时有分叉，顶裂片比侧裂片稍粗大，渐尖，或比侧裂片明显宽大，菱状椭圆形，长 2.2～2.7 cm，宽 0.3～0.7 cm，两侧具 2～4 对细裂齿或具 1 片小裂片，大孢子叶柄长 6～10.5 cm，径 0.4～0.6 cm，两侧着生胚珠 2～4（5）枚，扁球形，无毛，直径 0.3～0.5 cm，先端有小尖头，种子圆球形，长 1.7～2.7 cm，直径 1.6～2.4 cm，成熟时黄色、橘黄色至橘红色。

【花 果 期】花期 5 月，种期 10 月。

【分　　布】生长于云南个旧。中国特有。

[Scientific Name] *Cycas hongheensis*

[Morphological Characters] Tree 5-7 m tall，straight，sometimes branched，inflated toward base，30-40 cm in diameter，sometimes 1.3 m，bark gray，smooth，with longitudinal fissures，sometimes with protruding concentric ring，apex not tomentose，leaves base at the trunk. Scale-like leaves，lanceolate，stiff，(3-4.5) cm×(1-1.5) cm，apex acuminate，adaxial surface smooth，back surface densely yellowish-brown pubescent. Pinnate leaves 50-90（120）cm，V-shaped in cross section，leaflets in 49-58 pairs，opposite or near opposite，yellowish-brown pubescent when young，glabrescent，petiole 10-26 cm，with 6-22 pairs of short spines above the middle part，spines 0.1-0.2 cm，spines space 0.5-3.5 cm，middle leaflet (7-16) cm×(0.6-0.8) cm，margin flat or slightly curved，leathery，adaxially light green and glossy，yellowish-green abaxially，midvein adaxially flat，prominent abaxially；leaflets slightly warped，keel shaped，petiole and rachis yellowish-brown pubescent. Pollen cones cylindrical，18 cm×4.5 cm，yellowish-brown，microsporophylls cuneate，1.5-1.8 cm，apical sterile part peltate，0.8-1.1 cm in width，with densely short tomentose，apex acuminate；megasporophylls terminal lobes，ellipsoid to lanceolate，(4.5-8.5) cm×(3-5.5) cm，deeply divided into（6）10-16 pairs of fine subulate lobes on each side，subulate lobes，slender，middle lobes 1.5-2.5 cm in length，with base 1-2 cm in width，sometimes forked，terminal lobes a little longer than lateral lobes，with apex acuminate，some terminal lobes much longer than lateral lobes，rhombic ellipse，(2.2-2.7) cm×(0.3-0.7) cm，with fine lobes 2-4 pairs on both sides，or small lobes one pair，megasporophylls petiole (6-10.5) cm×(0.4-0.6) cm，ovules 2-4（or 5）on each side of stalk，oblate，glabrous，0.3-0.5 cm in diameter，apex acuminate. Seeds globose，(1.7-2.7) cm×(1.6-2.4) cm，yellow or orange when mature.

[Florescence and Fruit Period] Florescence May，seed October.

[Distribution] Gejiu in Yunnan Province. A native plant of China.

摄像：宋 鼎

Photo：Song Ding

19 多羽叉叶苏铁/ Duoyu chaye Sutie

【学　　名】*Cycas multifrondis*

【形态特征】亚地下茎，地上部分圆柱状，高 40 cm，直径 10～30 cm，叶痕宿存。羽叶 4～10 片，幼叶锈色，长 2.75～3.5 m，叶柄长 110～160 cm，下部疏被短柔毛，刺 40～78 对，刺长 3～5 cm，距 1.5～2.5 cm；羽片 54～88 枚，在叶轴近展；中部羽片长 22～41 cm，宽 1.4～2.5 cm，中脉两面隆起，条形，1～2 次二叉分歧，深绿色，有光泽，坚纸质至革质，先端渐尖，基部楔形，下侧明显下延，小叶柄长 0.5～3.5 cm。小孢子叶球纺锤状圆柱形，长 35～40 cm，小孢子叶楔形，不育部分盾状，宽 1.2～1.5 cm，密被短柔毛，先端具短尖头，长约 1 mm，两侧具 1～3 枚细齿。大孢子叶卵形，密被锈色绒毛，后逐渐脱落，边缘篦齿状深裂，两侧 16～19 对侧裂片，裂片纤细，先端芒尖，顶裂片钻形至披针形。种子近球形，直径约 3 cm，熟时黄褐色。

【花 果 期】花期 4—5 月，种期 10—11 月。

【分　　布】产于云南。中国特有。

【生　　境】常生长于海拔 100～1 000 m 的中低山石灰岩山地雨林下。

【保护价值】观赏。

【致危因素】盗挖散失，生境破碎化，种群过小。

【附　　注】缺少影像资料。

[Scientific Name] *Cycas multifrondis*

[Morphological Characters] Trunk almost subterranean，cylindrical above ground，40 cm×(10-30) cm，with leaf scars. Leaflets 4-10，2.75-3.5 m，rust when young，petiole 110-160 cm, with sparse，short tomentose at lower part，spines 40-78 pairs，3-5 cm in length，spacing 1.5-2.5 cm；leaflets 54-88，near rachis；middle leaflet (22-41) cm×(1.4-2.5) cm，midvein raised on both surfaces，liner，forked into 1 or 2 segments，dark green，glossy，stiff，papery to leathery，apex acuminate，base cuneate，decurrent，leaflets petiole 0.5-3.5 cm. Pollen cones spindle cylinder, 35-40 cm, microsporophylls cuneate, apical sterile part peltate, 1.2-1.5 cm in width, densely short tomentose, apex acuminate 1 mm, with fine lobes 1-3 on both sides. Megasporophylls ellipsoid, densely rust tomentose, glabrescent, deeply divided into 16-19 fine lobes on both sides, apex acuminate, terminal lobe subulate or lanceolate. Seeds subglobose, 3 cm in diameter, yellowish-brown when mature.

[Florescence and Fruit Period] Florescence April to May, seed October to November.

[Distribution] Yunnan Province. A native plant of China.

[Habitat] Limestone and rainforest in the middle and low mountains with an altitude of 100-1,000 m.

[Protection Value] Ornamental plant.

[Threatening Factors] Illegal digging, habitat fragmentation, population too small.

[Note] No image data.

20 多歧苏铁/ Duoqi Sutie

【学　　名】*Cycas multipinnata*

【形态特征】茎干约一半地下生，地上茎高 20～60 cm，径 15～25 cm；干皮黑褐色，具宿存叶痕。叶 1～2 片，三回羽裂，长 3～7 m，宽 70～150 cm；叶柄长 1.5～2.5 cm，粗 3～6 cm，每侧具刺 30～50 枚；一回裂片 12～22 对，与主轴夹角 60°～90°，羽片披针形，下部的最长，长 45～80 cm，宽 25～35 cm；二回裂片 5～13 枚，常互生，扇形或倒卵形，长 20～40 cm，宽 8～20 cm，在轴上斜展，3～7 次二叉分枝，具 1～5 cm 长的叶柄；三回裂片 1～3 次二叉分枝，不具柄或具短柄；羽片薄革质或纸质，倒卵状披针形或倒卵状线形，长 7～15 cm，宽 1.2～2.4 cm，先端尾状或尾状渐尖，尾尖长约 2 cm，基部下延常成翅，边缘波状，两面中脉隆起。孢子叶球近圆柱状，长 20～35 cm，金黄色，顶端钝圆；小孢子叶楔形，长 2～2.5 cm，宽 7～12 mm，先端钝圆，具短尖头，两侧具数枚小齿。大孢子叶长 8～15 cm，不育顶片三角状宽卵形，长 4～7 cm，宽 3～6.5 cm，裂片每侧 7～14 枚，钻状，长 1.5～3.5 cm；胚珠 6～10 枚。种子 6～10 枚，绿色，后变黄色，近球形，长 2.5～3.2 cm，中种皮具细乳突。孢子叶球期为 4—5 月。

【花 果 期】花期 4—5 月，种期 10—11 月。

【分　　布】产于云南。越南北部有分布。

【生　　境】生长于海拔 150～1 100 m 石灰岩山地茂密雨林下。

【保护价值】观赏。

【致危因素】生境退化和丧失；直接采挖；物种内在因素。

[Scientific Name] *Cycas multipinnata*

[Morphological Characters] Trunk half subterranean, 20-60 cm above ground, 15-25 cm in diameter; bark dark brown, with leaf scars. Leaves 1 or 2, 3-pinnate, 3-7 m in length, 70-150 cm; petiole (1.5-2.5) cm×(3-6) cm, with spines 30-50 on each side; pinnatifid leaves 12-22 pairs, longitudinally inserted at 60°-90° to primary rachis, leaflets lanceolate, longest lower part (45-80) cm×(25-35) cm; bipinnatifid leaves 5-13 pairs, often alternate, sector or obovate, (20-40) cm×(8-20) cm, at acute angle to rachis, forked into 3-7 segments, with petiole 1-5 cm; tripinnatifid leaves forked into 1-3 segments, short petiole or no petiole; leaflets leathery or papery, obovate-lanceolate or liner lanceolate, (7-15) cm×(1.2-2.4) cm, apex shortly acuminate to caudate, 2 cm, base decurrent, usually wing-like, margin wavy, midvein raised on both surfaces. Strobile cylindrical, 20-35 cm, golden yellow, apex blunt; microsporophylls cuneate, (2-2.5) cm×(7-12) mm, apex blunt, apex short acuminate, with lobes on both sides. Megasporophylls 8-15 cm, sterile blade triangular-ovate, (4-7) cm×(3-6.5) cm, deeply divided into 7-14 subulate lobes, 1.5-3.5 cm; ovules 6-10. Seeds 6-10, green, turning yellow, subglobose, 2.5-3.2 cm, sclerotesta finely verrucose. Strobile April to May.

[Florescence and Fruit Period] Florescence April to May, seed October to November.

[Distribution] Yunnan Province. Also found in North Vietnam.

[Habitat] Dense rainforest of limestone mountain 150-1,100 m above sea level.

[Protection Value] Ornamental plant.

[Threatening Factors] Habitat degradation and loss; direct digging; intrinsic factors of species.

摄像：朱鑫鑫　李西贝阳

Photo：Zhu Xinxin　Li Xibeiyang

21 攀枝花苏铁/ Panzhihua Sutie

【学　　名】*Cycas panzhihuaensis*

【形态特征】茎干圆柱状，高 1～2（3）m，直径 25～30 cm，顶端被厚绒毛；干皮呈暗褐或灰褐色，有宿存鳞状叶痕。叶 30～60 片，一回羽裂，长 0.7～1.3 m，宽 20～27 cm，蓝绿色；叶柄长 7～20 cm，在中上部有刺 5～13 对，基部密被褐色绒毛；羽片革质，蓝绿色，干时灰绿色，长 12～20 cm，宽 6～7 mm，基部下延，边缘平坦或稍反曲，中脉上面近平，下面隆起。小孢子叶球卵状圆柱形，长 25～45 cm；小孢子叶窄楔形，长 4～6 cm，宽 1.8～2 cm，先端具短尖。大孢子叶长 15～20 cm，密被具光泽淡褐色绒毛，后在裂片渐脱落，不育顶片菱状卵形，长 8～10 cm，宽 4～6 cm，裂片每侧约 20 枚，钻状，长 1～3 cm，顶生裂片更长；胚珠 4～6 枚，无毛。种子 2～4 枚，橘红色，球状或倒卵状球形，长 2.5～3.5 cm，外种皮肉质，干时变近膜质，脆易剥落，中种皮光滑。孢子叶球期为 4—5 月，种子在 9—10 月成熟。

【花 果 期】花期 3—4 月，种期 8—9 月。

【分　　布】产于四川、云南。中国特有。

【生　　境】常生长于海拔 1 100～2 000 m 干热河谷石灰岩山坡稀疏树灌丛。

【保护价值】观赏。

【致危因素】生境退化或丧失。

[Scientific Name] *Cycas panzhihuaensis*

[Morphological Characters] Trunk cylindric，1-2（3）m tall，25-30 cm in diameter，apex densely tomentose；bark dark brown to dark gray，scaly. Leaves 30-60，1-pinnate，(0.7-1.3)m×(20-27)cm，bluish green；petiole 7-20 cm，with 5-13 spines along each side of apical part，base densely brown tomentose；leaf blade leathery，bluish green，gray green when dry，(12-20) cm×(6-7) mm，base decurrent，margin flat or slightly recurved，midvein adaxially flat，raised abaxially. Pollen cones ellipsoid cylinder，25-45 cm；microsporophylls cuneate，(4-6) cm×(1.8-2) cm，apex acute. Megasporophylls 15-20 cm，brown tomentose，glabrescent，sterile blade triangular-ovate，(8-10) cm×(4-6) cm，deeply divided into 20 subulate lobes，1-3 cm in length，terminal lobes longer than lateral lobes；ovules 4-6，glabrous. Seeds 2-4，orange，globose or obovoid-globose，2.5-3.5 cm in length，exopleura fleshy，membranous when dry and brittle and easy to peel off，sclerotesta smooth. Strobile April to May，seed maturity September to October.

[Florescence and Fruit Period] Florescence March to April，seed August to September.

[Distribution] Sichuan Province，Yunnan Province. A native plant of China.

[Habitat] Thickets with sparse trees on limestone and sand shale of dry-hot valley，1,100-2,000 m above sea level.

[Protection Value] Ornamental plant.

[Threatening Factors] Habitat degradation or loss.

摄像：朱鑫鑫

Photo：Zhu Xinxin

22 篦齿苏铁/ Bichi Sutie

【学　　名】 *Cycas pectinate*

【形态特征】 树干圆柱形，高达 3 m。羽状叶长 1.2～1.5 m，叶轴横切面圆形或三角状圆形，柄长 15～30 cm，两侧有疏刺，刺略向下弯，长约 2 mm，羽状裂片 80～120 对，条形或披针状条形，厚革质，坚硬，直或微弯，边缘稍反曲，上部微渐窄，先端渐尖，基部窄，两侧不对称，下延生长，上面深绿色，中脉隆起，脉的中央常有一条凹槽，下面绿色，有散生短柔毛或渐变无毛，中脉显著隆起，中部的羽状裂片长 15～20 cm，宽 6～8 mm。雄球花长圆锥状圆柱形，长约 40 cm，径 10～15 cm，有短梗，小孢子叶楔形，长 3.5～4.5 cm，顶部三角状或斜方形，先端具钻形长尖头，宽 1.2～2 cm，密生褐黄色绒毛，花药 3～5（多为 4）个聚生；大孢子叶密被褐黄色绒毛，上部的顶片斜方状宽圆形或宽圆形，宽较长为大或长宽几相等，宽 6～8 cm，边缘有 30 余枚钻形裂片，裂片长 3～3.5 cm，先端尖，通常无毛，顶生的裂片较长大，长 4～5 cm，边缘常疏生锯齿或再分裂，大孢子叶的下部急缩成粗短的柄状，长 3～7 cm，上部两侧生胚珠 2～4 枚，胚珠无毛，卵圆形或近圆球形，顶端有一乳头状突起点。种子卵圆形或椭圆状倒卵圆形，长 4.5～5 cm，径 4～4.7 cm，熟时暗红褐色，具光泽，干后外种，皮常同中种皮分离开。

【花 果 期】花期 11 月至翌年 4 月，种期 9—10 月。

【分　　布】产于云南西南部，昆明有栽培。印度、尼泊尔、锡金、缅甸、泰国、柬埔寨、老挝、越南也有分布。

【生　　境】生长于海拔 800～1 800 m 的混交林中。

【致危因素】盗挖散失，生境破碎化。

[Scientific Name] *Cycas pectinate*

[Morphological Characters] Trunk cylindric，3 m in tall，pinnate leaves 1.2-1.5 m，rachis circular or triangular in cross section，petiole 15-30 cm，sparse and slightly recurved spines on both sides，2 mm，leaflets 80-120 pairs，liner or lanceolate，leathery，stiff，straight or slightly curved，apical slightly narrow，apex acute，base narrow，asymmetry，base decurrent，adaxially dark green，midvein raised，often a groove in the center of the vein，green abaxially，sparsely tomentose or glabrescent，midvein raised，middle leaflets 15-20 cm×6-8 mm. Pollen cones fusiform，40×10-15 cm，with short stalk，microsporophylls cuneate，3.5-4.5 cm，top triangle or rhombus，subulate long acuminate apex，1.2-2 cm in width，densely yellowish brown hairy，anther in groups of 3-5（often 4）；megasporophylls densely yellowish brown hairy，terminal lobe rhombus or circular，width equal to or longer than length，6-8 cm in width，with 30 lateral lobe subulate 3-3.5 cm，apex acuminate，usually glabrous，terminal lobe longer and larger than lateral lobes，4-5 cm，margin with sparsely lobes or again divided，megasporophylls base constricted into a short petiolule，3-7 cm，ovules 2-4 on each side of distal part of stalk，glabrous，ellipsoid or globose，a papillary process at tapex. Seeds ellipsoid or obovoid，4.5-5×4-4.7 cm，dark red when mature，glossy，exopleura often separate with sclerotesta when dry.

[Florescence and Fruit Period] Florescence November to next April，seed September to October.

[Distribution] Southwest Yunnan Province，cultivated in Kunming. Also found in India，Nepal，Sikkim，Myanmar，Thailand，Cambodia，Laos，Vietnam.

[Habitat] Mixed forest with altitude of 800-1,800 m.

[Threatening Factors] Illegal digging，habitat fragmentation.

摄像：林秦文

Photo：Lin Qinwen

23 苏铁/ Sutie

【学　　名】*Cycas revolute*

【形态特征】树干高约 2 m，稀达 8 m 或更高，圆柱形如有明显螺旋状排列的菱形叶柄残痕。羽状叶从茎的顶部生出，下层的向下弯，上层的斜上伸展，整个羽状叶的轮廓呈倒卵状狭披针形，长 75～200 cm，叶轴横切面四方状圆形，柄略成四角形，两侧有齿状刺，水平或略斜上伸展，刺长 2～3 mm；羽状裂片达 100 对以上，条形，厚革质，坚硬，长 9～18 cm，宽 4～6 mm，向上斜展微成 V 字形，边缘显著地向下反卷，上部微渐窄，先端有刺状尖头，基部窄，两侧不对称，下侧下延生长，上面深绿色有光泽，中央微凹，凹槽内有稍隆起的中脉，下面浅绿色，中脉显著隆起，两侧有疏柔毛或无毛。雄球花圆柱形，长 30～70 cm，径 8～15 cm，有短梗，小孢子飞叶窄楔形，长 3.5～6 cm，顶端宽平，其两角近圆形，宽 1.7～2.5 cm，有急尖头，尖头长约 5 mm，直立，下部渐窄，上面近于龙骨状，下面中肋及顶端密生黄褐色或灰黄色长绒毛，花药通常 3 个聚生；大孢子叶长 14～22 cm，密生淡黄色或淡灰黄色绒毛，上部的顶片卵形至长卵形，边缘羽状分裂，裂片 12～18 对，条状钻形，长 2.5～6 cm，先端有刺状尖头，胚珠 2～6 枚，生于大孢子叶柄的两侧，有绒毛。种子红褐色或橘红色，倒卵圆形或卵圆形，稍扁，长 2～4 cm，径 1.5～3 cm，密生灰黄色短绒毛，后渐脱落，中种皮木质，两侧有两条棱脊，上端无棱脊或棱脊不显著，顶端有尖头。

【花 果 期】花期 4—5 月，种期 9—10 月。

【分　　布】产于福建、台湾、广东，各地常有栽培。

【生　　境】生长于海拔 10～100 m 向阳近海边灌丛。

【致危因素】农垦及盗挖严重，20 世纪 70 年代在我国野生种群几乎绝迹。

[Scientific Name] *Cycas revolute*

[Morphological Characters] Trunk 2 m，rarely 8 m or taller，cylindrical，with spirally arranged rhombus petiole leaf scars. Pinnate leaves at the stem apex，lower layer downward curved，upper layer upward stretch，whole leaves obovoid or lanceolate，75-200 cm，petiole subtetragonal in cross section，flat or slightly oblique ascending with spines 2-3 mm；leaflets more than 100 pairs，liner，leathery，stiff，9-18 cm×4-6 mm，lightly V-shaped obliquely upward，margin strongly recurved，apical slightly narrow，with apex acuminate，narrow base，asymmetrical on two sides，base decurrent，adaxially dark green and glossy，slightly depressed in the center，with raised midvein in the groove，light green caudally，with midvein strongly raised，sparsely tomentose or no tomentose on both sides. Pollen cones cylindric，30-70×8-15 cm，with short stalk，microsporophylls narrowly cuneate，3.5-6×1.7-2.5 cm，top rounded-truncate，with acuminate apex 5 mm，straight，narrow caudally，apical carinate，with midrib and top densely yellowish-brown or gray-yellow long tomentose，anther usually 3 in groups; megasporophylls 14-22 cm, densely light yellow or gray yellow tomentose, terminal lobe ellipsoid or long ellipsoid, margin deeply divided into 12-18 pairs of liner lobes, 2.5-6 cm, with apex acuminate, ovules 2-6 on each side of stalk, tomentose. Seeds reddish brown or orange, obovoid or ellipsoid, somewhat compressed, 2-4×1.5-3 cm, densely gray yellow short tomentose, glabrescent, sclerotesta woody with two ridges on both sides, no obvious apical ridge, with apex acuminate.

[Florescence and Fruit Period] Florescence April to May, seed September to October.

[Distribution] Fujian Province, Taiwan Province, Guangdong Province, cultivated everywhere.

[Habitat] Thickets exposed to the sun and near the sea, 10-100 m above sea level.

[Threatening Factors] Reclamation and illegal digging. In the 1970s, its wild population almost disappeared in China.

摄像：林秦文

Photo: Lin Qinwen

24 叉孢苏铁/ Chabao Sutie

【学　　名】*Cycas segmentifida*

【形态特征】茎干约一半地下生，地上茎高 15～70 cm，直径 10～40 cm；干皮黑褐色，具宿存叶痕。叶 15～25 片，一回羽裂，长 2～3.3 m，宽 45～60 cm；叶柄长 60～150 cm，初呈蓝绿色，后变绿色，具刺 25～50 对，刺距 1～2.5 cm，刺长（1）2～3.5 mm；羽片间距 1～1.8 cm，薄革质，长 21～40 cm，宽 12～18 mm，基部近对称，稍下延，边缘有时波状，中脉鲜时两面隆起，干时下面变平或稍凹。小孢子球窄椭圆状圆柱形，长 30～50 cm；小孢子叶窄楔形，长 2～2.5 cm，宽 1～1.2 cm，先端具短尖头。大孢子叶 25～50 片，密集，长 10～16（20）cm，不育顶片宽卵形或心状卵形，长 5～9 cm，宽 4～8（11）cm，裂片钻状或丝状，每侧 16～22 枚，其中部分呈二叉状分裂，长 2～5 cm，顶生裂片钻状或披针形，边缘常具不整齐的齿；胚珠（2）3～6 枚，无毛。种子倒卵状，长 2.8～3.5 cm，基部窄楔形，中种皮具细疣状突起。孢子叶球期为 3—6 月，种子在 11—12 月成熟。

【花 果 期】花期 4—5 月，种期 9—10 月。

【分　　布】产于广西，贵州望谟、册亨，云南富宁。中国特有。

【生　　境】生长于海拔 350～800 m 石灰石、页岩、片岩基质土壤的河谷阔叶林下。

【致危因素】盗挖散失，农垦使生境破碎化。

[Scientific Name] *Cycas segmentifida*

[Morphological Characters] Trunk half subterranean, 15-70×10-40 cm above ground; bark dark brown, with leaf scars. Leaves 15-25, 1-pinnate, 2-3.3 m×45-60 cm; petiole blue in 1st year, later green, 0.6-1.5 m, with 25-50 spines (1) 2-3.5 mm along each side, 1-2.5 cm apart; leaf blade 1-1.8 cm apart, leathery, 21-40×1.2-1.8 cm, base symmetrical, slightly decurrent, margin flat or somewhat undulate, midvein raised on both surfaces only when fresh, flat abaxially when dry. Pollen cones narrowly ellipsoid-cylindric, 30-50 cm; microsporophylls narrowly cuneate, 2-2.5×1-1.2 cm, apex with short mucro. Megasporophylls 25-50, densely grouped, 10-16(20)cm, sterile blade broadly ovate or cordate-ovate, 5-9×4-8 (11) cm, with lobe subulate or filiform, 16-22 on each side, some forked lobes 2-5 cm, terminal lobe subulate to rhombic-lanceolate, margin pectinate; ovules(2)3-6, glabrous. Seeds obovoid, 2.8-3.5 cm, base contracted into point, sclerotesta finely verrucose. Strobile Mar-Jun, seed maturity Nov-Dec.

[Florescence and Fruit Period] Florescence April to May, seed period September to October.

[Distribution] Guangxi Zhuang Autonomous Region, Wangmo and Ceheng in Guizhou Province, Funing in Yunnan. A native plant of China.

[Habitat] Valley broad-leaved forest with limestone, shale and schist matrix soil at an altitude of 350-800 m.

[Protection Value] Ornamental plant.

[Threatening Factors] Illegal digging and reclamation makes habitat fragmented.

摄像：黄江华　宋　鼎　华国军

Photo: Huang Jianghua　Song Ding　Hua Guojun

25 石山苏铁/ Shishan Sutie

【学　　名】*Cycas sexseminifera*

【形态特征】石山苏铁为小型灌木，树干通常不明显，有时也膨大呈葫芦状，或纺锤状，或盘状，或圆柱形，高可达 50 cm，直径达 25 cm，基部膨大成圆球形，灰色至灰褐色，叶痕宿存，后期常脱落而光滑，无茎顶绒毛；鳞叶披针形，长 4.5～8.5 cm，宽 1.5～2.5 cm，暗棕色，背面密被短绒毛；羽叶长（30）50～170 cm，叶柄长 5～63 cm，上部两侧具 3～35 对短刺，刺长 0.1～0.4 cm，间距 0.6～2.2 cm，羽片 40～81 对，水平开展，革质，羽片条形，中部羽片长 7.5～28（40）cm，宽 0.5～1.2 cm，先端渐尖，具短尖头，基部不对称，下侧下延生长，中脉上面平或微隆起，下面明显隆起，上面深绿色，有亮泽，下面淡绿色，叶边缘平或有时反卷，叶轴、叶柄及叶背密被锈色柔毛。

【花 果 期】花期 4—6 月，种期 9—11 月。

【分　　布】产于广西。越南中北部有分布。

【生　　境】生长于海拔 200～500 m 石灰岩山地的裸露石缝中。

【保护价值】观赏。

【致危因素】盗挖散失，生境破碎化。

[Scientific Name] *Cycas sexseminifera*

[Morphological Characters] Small shrub，trunk generally subterranean，50 cm tall，25 cm in diameter，sometimes swollen and gourd or spindle shaped or cylindrical，bark gray to gray-brown，with leaf scars，nearly smooth toward base of trunk，glabrous apically；scale-like leaves lanceolate，4.5-8.5×1.5-2.5 cm，dark brown，densely short tomentose abaxially；pinnate leaves（30）50-170 cm，petiole 5-63 cm，with 3-35 spines along each side of apical part，spines 0.1-0.4 cm，0.6-2.2 cm apart，leaflets 40-81 pairs，flat，leathery，liner，median leaflets 7.8-28（40）×0.5-1.2 cm，apex with short mucro，base asymmetric and decurrent，midvein nearly flat or slightly raised，strongly raised abaxially，adaxially dark green and glossy，light green abaxially，margin flat or only slightly revolute，rachis，petiole and underneath rust tomentose.

[Florescence and Fruit Period] Florescence April to June，seed September to November.

[Distribution] Guangxi Zhuang Autonomous Region. Also found in North Central Vietnam.

[Habitat] In the exposed crevices of limestone mountains 200-500 m above sea level.

[Protection Value] Ornamental plant.

[Threatening Factors] Illegal digging，habitat fragmentation.

摄像：林秦文

Photo：Lin Qinwen

26 单羽苏铁/ Danyu Sutie

【学　　名】*Cycas simplicipinna*

【形态特征】单羽苏铁为低矮灌木，主干不明显，叶痕宿存。鳞叶披针形，长 4.5～7 cm，宽 1.5～2 cm，羽叶长（1）1.5～2.5 m，叶柄长 0.2～1 m，刺 19～39 对，长约 0.3 cm，刺间距 1.5～4 cm，羽片 18～81 对，中部羽片条形，长 17～40 cm，宽（1.2）1.6～2.5 cm，深绿色，有光泽，纸质至薄革质，先端渐尖，中脉长 1～1.4 cm，宽达 1～1.2 cm，顶端近截状，外面被锈色柔毛。大孢子叶两面隆起。小孢子叶球狭长圆柱形，长 15～21 cm，直径 2～4 cm，小孢子叶楔形，长 16 cm，顶片近菱形至卵形，长 3.5～6 cm，宽 3～5 cm，边缘篦齿状深裂，两侧具 5～11 对侧裂片，裂片长 1～2 cm，顶裂片钻形至三角状披针形，有时具浅齿，大孢子叶柄部被毛，胚珠 2～5 枚。种子为椭圆形，长 2.5～2.7 cm，直径 2 cm。

【花 果 期】花期 5—6 月，种期 10 —11 月。

【分　　布】产于云南。泰国北部、越南、老挝、缅甸有分布。

【生　　境】生长于海拔 500～800 m 的阴湿热带山坡常绿林下。

【保护价值】观赏。

【致危因素】生境破碎化，盗挖散失，种群过小。

【附　　注】缺少影像资料。

[Scientific Name] *Cycas simplicipinna*

[Morphological Characters] Low shrub，trunk subterranean，with leaf scars，4.5-7×1.5-2 cm，pinnate leaves（1）1.5-2.5 m，petiole 0.2-1 m，with spines 19-39 pairs，0.3 cm，1.5-4 cm apart，leaflets 18-81 pairs，median leaflet liner，17-40×（1.2）1.6-2.5 cm，dark green and glossy，papery to leathery，with acuminate apex，midvein 1-1.4×1-1.2 cm，apex truncate，with rust tomentose. Megasporophylls raised on both surfaces. Pollen cones liner cylindric，15-21×2-4 cm，microsporophylls cuneate，16 cm，terminal lobe ellipsoid，

3.5-6×3-5 cm，margin pectinate，deeply divided into lobes 5-11 pairs subulate，1-2 cm，with terminal lobe subulate to triangle- lanceolate，sometimes slightly pectinate，megasporophylls petiole tomentose， with ovules 2-5. Seeds ellipsoid， 2.5-2.7×2 cm.

[Florescence and Fruit Period] Florescence May to June，seed October to November.

[Distribution] Yunnan Province. Also found in Northern Thailand，Vietnam，Laos，Myanmar.

[Habitat] Evergreen forest on the humid tropical hillside 500-800 m above sea level.

[Protection Value] Ornamental plant.

[Threatening Factors] Habitat fragmentation， illegal digging， population too small.

[Note] No image data.

27 四川苏铁/ Sichuan Sutie

【学　　名】*Cycas szechuanensis*

【形态特征】树干圆柱形，直或弯曲，高 2～5 m。羽状叶长 1～3 m，集生于树干顶部；羽状裂片条形或披针状条形，微弯曲，厚革质，长 18～34 cm，宽 1.2～1.4 cm，边缘微卷曲，上部渐窄，先端渐尖，基部不等宽，两侧不对称，上侧较窄，几靠中脉，下侧较宽、下延生长，两面中脉隆起，上面深绿色，有光泽，下面绿色。大孢子叶扁平，有黄褐色或褐红色绒毛，后渐脱落，上部的顶片倒卵形或长卵形，长 9～11 cm，宽 4.5～9 cm，先端圆形，边缘篦齿状分裂，裂片钻形，长 2～6 cm，粗约 3 mm，先端具刺状长尖头，无毛，下部柄状，长 10～12 cm，密被绒毛，下部的绒毛后渐脱落，在其中上部每边着生 2～5（多为 3～4）枚胚珠，上部的 1～3 枚胚珠的外侧常有钻形裂片生出，胚珠无毛。

【花 果 期】花期 4—6 月。

【分　　布】四川西部峨眉山、乐山、雅安及福建南平等地有栽培。

【生　　境】生长于海拔 300～500 m 的阴湿林中。

【保护价值】观赏。

【致危因素】农垦使生境破碎化，盗挖散失，种群过小。

[Scientific Name] *Cycas szechuanensis*

[Morphological Characters] Trunk cylindrical, straight or bending, 2-5 m tall. Pinnate leaves 1-3 m, at trunk apex in groups; leaflets liner or lanceolate, slightly curved, leathery, 18-34×1.2-1.4 cm, margin slightly recurved, apex acuminate, base asymmetric on both surface, narrow apically, near midvein, broad caudally, base decurrent, midvein strongly raised on both surfaces, adaxially dark green and glossy, green abaxially. Megasporophylls flat, yellowish brown or reddish brown tomentose, glabrescent, terminal lobe obovate, or ovate, 9-11×4.5-9 cm, round apically, margin laciniate, with subulate lobes 2-6 cm×3 mm, apex with long mucro, glabrous, base constricted into a short petiolule 10-12 cm, densely tomentose, glabrescent, ovules 2-5 (usually 3-4) on each side of apical stalk, with subulate lobes on the upper 1-3 ovules, ovules glabrous.

[Florescence and Fruit Period] Florescence April to June.

[Distribution] Mount E'mei, Mount Le, Ya'an in East Sichuan Province, Nanping in Fujian Province.

[Habitat] Dark and damp forest, 300-500 m above the sea level.

[Protection Value] Ornamental plant.

[Threatening Factors] The cultivation makes the habitat fragmented, illegal digging, and the population is too small.

摄像：华国军　李晓东　徐克学

Photo: Hua Guojun　Li Xiaodong　Xu Kexue

28 台东苏铁/ Taidong Sutie

【学　　名】*Cycas taitungensis*

【形态特征】茎干圆柱状，高达 5 m，直径达 45 cm，顶端密被厚绒毛；干皮黑褐色，具宿存叶痕。叶约 50 片，一回羽裂，长 1.3～2 m，宽 20～40 cm；叶柄长 15～30 cm，具刺 7～14 对；羽片直或近镰刀状，革质，长 14～20 cm，宽 5～8 mm，基部下延，先端渐窄，下面疏被柔毛，边缘平或稍反曲，中脉上面绿色，近平坦，下面淡绿，显著隆起，横断面呈宽 V 字形。小孢子叶球卵状圆柱形，长 45～55 cm；小孢子叶窄倒三角形，长 3.5～4.5 cm，宽 1.1～1.5 cm，先端具骤尖头。大孢子叶橘红色，长 15～25 cm，被淡褐色绒毛，后渐脱落，不育顶片菱状圆形或近圆形，长 7～14 cm，宽 6～11 cm，边缘半裂，裂片每侧 14～19 枚，钻状，长 2～4.5 cm；胚珠 4～6 枚，密被褐色绒毛。种子 2～6 枚，橘红色，干时带紫红色，窄倒卵状或椭圆状，长 4～5 cm，常疏被残存的绒毛，中种皮两侧常具不规则的 2 或 3 沟。孢子叶球期为 4—6 月，种子在 9—10 月成熟。

【花 果 期】花期 4—6 月，种期 9—10 月。

【分　　布】产于台湾。中国特有。

【生　　境】生长于海拔 300～950 m 陡峭的山坡林中。

【保护价值】观赏。

【致危因素】生境破碎化，盗挖散失，近年病虫害严重，种群过小。

[Scientific Name] *Cycas taitungensis*

[Morphological Characters] Trunk cylindrical，5 m tall，45 cm in diameter，apex very tomentose；bark dark brown，scally. Leaves 50，1-pinnate，1.3-2 m×20-40 cm；petiole 15-30 cm，with 7-14 pairs of spines；leaflets straight of subfalcate，leathery，14-20 cm×5-8 mm，base decurrent，apex attenuate，sparsely pubescent abaxially，margin flat or slightly recurved，midvein adaxially green and flat，abaxially light green and raised strongly，broadly V-shaped in cross section. Pollen cones ovoid-cylindric，45-55 cm；microsporophylls narrowly obtriangular，3.5-4.5×1.1-1.5 cm，apex cuspidate. Megasporophylls bright orange-red，15-25 cm，pale brown tomentose，glabrescent，sterile blade suborbicular or rhombic-orbicular，7-14×6-11 cm，pectinate，with 14-19 subulate lobes 2-4.5 cm on each side；ovules 4-6，densely pale brown tomentose. Seeds 2-6，orange-red，or purplish red when dry，narrowly obovoid or subellipsoid，4-5 cm，often with remnant hairs，sclerotesta with 2 or 3 irregular grooves on each side. Pollination Apr-Jun，seed maturity Sept-Oct.

[Florescence and Fruit Period] Florescence April to June，seed September to October.

[Distribution] Taiwan. A native plant of China.

[Habitat] Steep hillside forest at an altitude of 300-950 m.

[Protection Value] Ornamental plant.

[Threatening Factors] Habitat fragmentation，illegal digging，serious diseases and insect pests in recent years，too small population.

摄像：徐晔春　吴棣飞　徐锦泉

Photo：Xu Yechun　Wu Difei　Xu Jinquan

29 闽南苏铁（台湾苏铁）/ Minnan Sutie（Taiwan Sutie）

【学　　名】*Cycas taiwaniana*

【形态特征】茎干圆柱状，高 0.2～3.5 m，直径 15～35 cm；干皮黑褐色，有宿存叶痕。叶 50～80 片，一回羽裂，长 1.5～3 m，宽 40～60 cm；叶柄长 40～120 cm，具刺 30～60 对，刺距 1～2.5 cm，基部常无刺；羽片与叶轴夹角 55°～75°，常镰刀状，薄革质或坚纸质，长 18～35 cm，宽 11～14 mm，基部稍下延，边缘有时微波状，中脉湿时两面隆起，干时下面变平或微凹。鳞叶先端具硬尖头。小孢子叶球卵状或椭圆状圆柱形，长 30～45 cm；小孢子叶窄楔形，长 2～3 cm，宽 5～8 mm，先端具短尖头。大孢子叶长 15～22 cm，被淡褐色绒毛，不育顶片菱状卵形、卵形或心状卵形，长 7～12 cm，宽 6～7 cm，篦齿状深裂或半裂，裂片每侧 11～23 枚，钻状，长 2～3.5 cm，顶生裂片钻状或稍扁化，长 2.5～4 cm；胚珠 4～6 枚，无毛。种子 2～4 枚，淡黄色，近球状，长 2.8～3.3 cm，中种皮具细疣状突起。孢子叶球期为 4—5 月，种子在 9—10 月成熟。

【花 果 期】花期 4—5 月，果期 9—10 月。

【分　　布】产于福建诏安、广东汕头。中国特有。

【生　　境】生长于海拔 200～400 m 的向阳丘陵土壤。

【保护价值】观赏。

【致危因素】生境破碎化，盗挖散失，种群过小。

【附　　注】缺少影像资料。

[Scientific Name] *Cycas taiwaniana*

[Morphological Characters] Trunk cylindric，0.2-3.5 m×15-35 cm；bark dark brown，scaly. Leaves 50-80，1-pinnate，1.5-3 m×40-60 cm；petiole 40-120 cm，with 30-60 spines along each side of basal part，1-2.5 cm apart，base glabrous；leaflets longitudinally inserted at 55°-75° to rachis，often falcate，leathery or stiff and papery，18-35 cm×11-14 mm，base decurrent，margin sometimes slightly wavy，midvein raised on both surfaces when fresh，flat or slightly concave when dry. Scale-like leaves apex acuminate. Pollen cones ovoid or ellipsoid-cylindric, 30-45 cm; microsporophylls narrowly cuneate, 2-3 cm×5-8 mm, with very short, upcurved mucro. Megasporophylls 15-22 cm, pale brown tomentose, sterile blade rhombic-ovate, ovate or heart-ovate, 7-12×6-7 cm, margin pectinate, with 11-23 lobes 2-3.5 cm, terminal lobe subulate or flattened, 2.5-4 cm; ovules 4-6, glabrous. Seeds 2-4, pale yellowish, globose, 2.8-3.3 cm, sclerotesta finely verrucose. Pollination Apr-May, seed maturity Sept-Oct.

[Florescence and Fruit Period] Florescence April to May, fruit September to October.

[Distribution] Zhao'an in Fujian Province, Shantou in Guangdong Province. A native plant of China.

[Habitat] Sunny hilly soil 200-400 m above sea level.

[Protection Value] Ornamental plant.

[Threatening Factors] Habitat fragmentation, illegal digging, population too small.

[Note] No image data.

30 绿春苏铁/ Lyuchun Sutie

【学　　名】*Cycas tanqingii*

【形态特征】亚地下茎，地上部分圆柱状，高达 2 m，无茎顶绒毛，4～9 片羽叶集生茎顶。羽叶深绿色，有光泽，长 1.9～3.6 m，平展，具 100～120 枚小羽片，叶轴顶端小羽片成对；叶柄全部具刺，长 70～160 cm；中部小羽片明显两面不同色，上面深绿色，有光泽，下面苍绿色；条形，长 30～40 cm，宽 1.5～2.2 cm。与叶轴成 65°～85°，边缘轻度反曲，中脉两面隆起。鳞叶狭三角形，长 5～8 cm，柔软多毛，宿存。小孢子叶球大，纺锤形，黄色至橘黄色，长 40 cm；孢子叶柔软，长 2.5～3 cm，不育顶点长 2～3 mm，水平至隆起，顶刺突出，急剧反卷。大孢子叶球包被紧密，近球形；大孢子叶不育顶片菱状阔圆形，背部被褐色绒毛，掌状深裂，侧裂片每侧 6～9 枚，长 1.5～4 cm，顶裂片与侧裂片不同或相似。胚珠 2 枚，种子大，倒卵状球形，长 3.5～4 cm，肉质种皮黄色，无纤维层，硬种皮具疣状突起，无海绵状内种皮。

【花 果 期】花期 4—5 月，种期 9—10 月。

【分　　布】产于云南绿春。越南北部有分布。

【生　　境】生长于海拔 500～800 m 的雨林内。

【保护价值】观赏。

【致危因素】生境破碎化，盗挖散失。

[Scientific Name] *Cycas tanqingii*

[Morphological Characters] Trunk subterranean，cylindrical above ground，2 m tall，apex glabrous，with 4-9 pinnate leaves in group. Pinnate leaves dark green，glossy，1.9-3.6 m，flat，with 100-120 leaflets，rachis apex with leaflets in pairs；spines all over the petiole，

70-160 cm; the two surfaces of median leaflets in different color, dark green apically, pale green abaxially; liner, 30-40 cm×1.5-2.2 cm. Longitudinally inserted at 65°-85° to rachis, midvein raised on both surfaces. Scale-like leaves, narrowly triangle, 5-8 cm, soft and tomentose, with leaf scars. Pollen cones large, fusiform, yellow to orange, 40 cm; microsporophylls soft, 2.5-3 cm, apical sterile part 2-3 cm, flat to raised, apex with sharp mucro, strongly recurved. Megasporophylls covered, subglobose; apical sterile part subulate-round, brown tomentose abaxially, deeply divided into 6-9 lobes on each side, 1.5- 4 cm, terminal lobe different from or close to lateral lobes. Ovules 2; seeds large, obovoid, 3.5-4 cm, succulent, yellow, no fibrous layer, sclerotesta verrucose, without spongy endopleura.

[Florescence and Fruit Period] Florescence April to May, seed September to October.

[Distribution] Lyuchun in Yunnan Province. Also found in North Vietnam.

[Habitat] Rain forest with an altitude of 500-800 m.

[Protection Value] Ornamental plant.

[Threatening Factors] Habitat fragmentation, illegal digging.

摄像：林秦文

Photo: Lin Qinwen

31 银杏/ Yinxing

【学　　名】 *Ginkgo biloba*

【形态特征】 乔木，高达 40 m，胸径可达 4 m；幼树树皮浅纵裂，大树之皮呈灰褐色，深纵裂，粗糙；幼年及壮年树冠圆锥形，老则广卵形；枝近轮生，斜上伸展（雌株的大枝常较雄株开展）；一年生的长枝淡褐黄色，二年生以上变为灰色，并有细纵裂纹；短枝密被叶痕，黑灰色，短枝上亦可长出长枝；冬芽黄褐色，常为卵圆形，先端钝尖。叶扇形，有长柄，淡绿色，无毛，有多数叉状并列细脉，顶端宽 5～8 cm，在短枝上常具波状缺刻，在长枝上常 2 裂，基部宽楔形，柄长 3～10（多为 5～8）cm，幼树及萌生枝上的叶常较而深裂（叶片长达 13 cm，宽 15 cm），有时裂片再分裂（这与较原始的化石种类之叶相似），叶在一年生长枝上螺旋状散生，在短枝上 3～8 叶呈簇生状，秋季落叶前变为黄色。球花雌雄异株，单性，生长于短枝顶端的鳞片状叶的腋内，呈簇生状；雄球花葇荑花序状，下垂，雄蕊排列疏松，具短梗，花药常 2 个，长椭圆形，药室纵裂，药隔不发；雌球花具长梗，梗端常分两叉，

稀 3～5 叉或不分叉，每叉顶生一盘状珠座，胚珠着生其上，通常仅一个叉端的胚珠发育成种子，内媒传粉。种子具长梗，下垂，常为椭圆形、长倒卵形、卵圆形或近圆球形，长 2.5～3.5 cm，径为 2 cm，外种皮肉质，熟时黄色或橙黄色，外被白粉，有臭叶；中处皮白色，骨质，具 2～3 条纵脊；内种皮膜质，淡红褐色；胚乳肉质，味甘略苦；子叶 2 枚，稀 3 枚，发芽时不出土，初生叶 2～5 片，宽条形，长约 5 mm，宽约 2 mm，先端微凹，第 4 或第 5 片起之后生叶扇形，先端具一深裂及不规则的波状缺刻，叶柄长 0.9～2.5 cm；有主根。

【花 果 期】花期 4 月，种子 9 —10 月成熟。

【分　　布】银杏的栽培区甚广：北自东北沈阳，南达广州，东起华东，西南至贵州、云南西部（腾冲）。

【生　　境】野生状态的银杏生长于针阔叶林带。

【保护价值】观赏、行道树。

【致危因素】冰川影响，为残存的“活化石”植物。

[Scientific Name] *Ginkgo biloba*

[Morphological Characters] Trees to 40 m tall; trunk to 4 m d.b.h.; bark slightly and longitudinally fissured when young, grayish brown when mature, deeply and longitudinally fissured, rough; crown conical when young and mature, broadly ovoid when old; branchlets slant upward (main branches of female plant more extensive than male); long branches pale brown in 1st year, turning gray in 2nd year, longitudinally fissured; short branchlets blackish gray, with dense, irregularly elliptic leaf scars, sometimes have long branches on short branches; winter buds yellowish brown, ovate, apex with blunt mucro. Leaves fan shaped, with long petiole, pale green, glabrous, forked veins on surface, apex 5-8 cm wide, those on short branchlets with undulate distal and margin notched apex, those on long branchlets divided by a deep, apical sinus into 2 lobes each further dissected, leaves blade base broadly cuneate, petiole 3-10 cm (usually 5-8 cm), leaves of young trees deeply fissured (leaf blade 13×15 cm), sometimes divided again (similar to the leaves of the more primitive fossil species), leaves spirally scattered on branches the 1st year, 3-8 in groups on short branches, turning yellow in autumn. Cones dioecious, unisexual, axils of scale-like leaves in groups, on top of short branches; male cones catkin, dropping; female cones sparse, with short petiole, pollen cones usually 2, elliptic, with gaping slit, connective glabrous; female cones with long petiole, apex usually forked into 3-5 lobes or complete, with discoid ovule base, usually only one ovule on one forked apex develops in to seed, self-pollination. Seeds with long petiole, dropping, narrowly obovoid, ovoid, or subglobose, 2.5×3.5 cm in length, 2 cm in diameter, sarcotesta yellow when mature, or orange-yellow, glaucous, with rancid odor when ripe; sclerotesta white, with 2 or 3 longitudinal ridges; endotesta pale reddish brown; endosperm succulent, tastes sweet and little

bitter; cotyledons 2, rarely 3, not unearthed during germination; primary leaves 2-5, broad stripe, 5×2 mm, apex slightly concave, leaves turning fan shaped after the 4th or 5th leaf, apex with deeply gaping slit, apex with deep gap and undulate cut, petiole 0.9-2.5 cm; with main root.

[Florescence and Fruit Period] Florescence April, seed September to October.

[Distribution] It is cultivated in a wide range of areas: from Shenyang in the northeast to Guangzhou in the south, East China in the East, Guizhou Province and Western Yunnan Province (Tengchong) in the southwest.

[Habitat] Ginkgo biloba in wild state grows in coniferous and broad-leaved forest belt.

[Protection Value] Ornamental plant, border tree.

[Threatening Factors] Affected by glaciers, the remaining "living fossil" plants.

摄像：林秦文

Photo: Lin Qinwen

32 百山祖冷杉/ Baishanzu Lengshan

【学　　名】*Abies beshanzuensis*

【形态特征】乔木，高约 11 m，胸径 40 cm；树皮灰白色，不规则龟裂，裂块纵向大于横向，裂隙较深；大枝平展，枝皮不规则浅裂；小枝对生，稀三枝轮生，基部围有宿存芽鳞，主干及直立枝上的小枝交叉对生，一年生枝淡黄色或黄灰色，无毛或凹槽中有疏毛；冬芽卵圆形，有树脂，生长于枝顶，三个排成一平面，中间之芽常较两侧之芽为大，稀在小枝节间上面（即背地面）生有一芽、其后发育成直立枝，芽鳞淡黄褐色，三角状卵形，背面中上部具钝纵脊。叶条形，在侧枝上排列成二列状，或枝条下面之叶排列成两列，上面之叶斜展至直伸，其长度由下面两侧至上面中央递减，长 1～4.2（通常为 1.5～

3.5）cm，宽 2.5～3.5 mm，先端有凹缺，下面有两条白色气孔带；横切面有 2 个边生树脂道或生长于两侧端的叶肉薄壁组织内，上面至下面两端及下面中部有一层连续排列的皮下层细胞；幼枝之叶在枝上常呈两列，先端呈二尖裂。雌球花圆柱形，长 3～3.5 cm，径约 8 mm，苞鳞上部向后反曲。球果通常每一枝节之间着生 1～3 个，圆柱形，长 7～12 cm，径 3.5～4 cm，成熟前绿色至淡黄绿色，熟后淡褐黄色或淡褐色；中部种鳞扇状四边形，稀近肾状四边形，长 1.8～2.4 cm，宽 2.5～3 cm，先端近全缘或有极细之细齿，两侧边缘有不规则锯齿，基部楔形、两侧耳状；苞鳞稍短于种鳞或几相等长，上部近圆形，宽 7～8 mm，边缘有细齿，先端露出、反曲，尖头短，长 1 mm 以内，中部收缩或窄缩呈条状，宽 3.5～4 mm；种子倒三角状，具与种子等长而宽大的膜质种翅，翅端平截。

【花 果 期】花期 5 月，球果 11 月成熟。

【分　　布】产于浙江庆元，目前仅存三株大树保留，亟待加强保护和繁殖。中国特有。

【生　　境】生长于海拔约 1 700 m 的高山针阔混交林中。

【保护价值】保护残存植株、科研。

【致危因素】冰川气候影响。难以授粉，种子很难成熟和长成幼树。

[Scientific Name] *Abies beshanzuensis*

[Morphological Characters] Trees to 11 m tall；trunk to 40 cm d.b.h.；bark pale gray，with irregularly deep gaps，longitudinal gap larger than the transverse；branches spreading，irregularly scaly；branchlets in group of two，rarely in group of three，base scaly，branchlets on the main and vertical branches cross each other，branchlets initially yellow or gray-yellow in 1st year，glabrous or puberulent；winter buds on branch apex，ovoid resinous，three in a plane，with mid-bud larger than the other two，rarely a bud on branch adaxial，growing to be straight branch. Bud scale pale yellow-brown，triangular ovate，dorsal middle upper part with blunt longitudinal ridge. Liner leaves，pectinately arranged in 2 lateral sets on lower side and on lateral branchlets，and stretching on upper side，length decreases from the bottom two sides to the top center，1-4.2 cm（usually 1.5-3.5 cm）×2.5-3.5 mm，apex emarginate，stomatal lines in 2 white bands abaxially；resin canals 2 on cross section or in parenchyma of mesophyll at both ends，continuous layer of subcutaneous cells at both surfaces and the middle part

abaxially; leaves pectinately arranged in 2 on young branches, apex forked. Female cones cylindric, 3-3.5 cm×8 mm, scales backward curve. Seed cones 1-3 at each joint, cylindric, 7-12×3.5-4 cm, green or pale yellow-green before mature, turning pale yellow-brown or pale brown; seed scales at middle of cones flabellate-trapeziform, rarely reniform-trapeziform, 1.8-2.4×2.5-3 cm, apex and margin erose-denticulate laterally, base cuneate, lateral auriculate; bud scales shorter than seed scales, or equal in length, distal part orbicular, 7-8 mm in width, margin erose-denticulate laterally, apex cuspidate, cusp exserted, reflexed, acute, within 1 mm, constricted at middle, 3.5-4 mm in width; seeds obtriangular, with broadly dolabriform wing the same length as seeds.

[Florescence and Fruit Period] Florescence May，seed cones mature in November.

[Distribution] Qingyuan in Zhejiang Province，only three remaining trees remain. It is urgent to strengthen protection and reproduction. A native plant of China.

[Habitat] About 1,700 m high mountain coniferous and broad-leaved mixed forest.

[Protection Value] Protection of remaining plants，scientific research.

[Threatening Factors] Glacial climate impact. It’s difficult for pollination，seed to mature and young trees to grow.

摄像：武　晶　曾玉亮　刘　军　吴棣飞

Photo：Wu Jing　Zeng Yuliang　Liu Jun　Wu Difei

33 秦岭冷杉/ Qin Ling Lengshan

【学　　名】 *Abies chensiensis*

【形态特征】 乔木，高达 50 m；一年生枝淡黄灰色、淡黄色或淡褐黄色，无毛或凹槽中有稀疏细毛，二、三年生枝淡黄灰色或灰色；冬芽圆锥形，有树脂。叶在枝上列成两列或近两列状，条形，长 1.5～4.8 cm，上面深绿色，下面有 2 条白色气孔带；果枝之叶先端尖，或钝，树脂道中生或近中生，营养枝及幼树的叶较长，先端二裂或微凹，树脂管边生；横切面上面至下面两侧边缘有皮下细胞一层，连续或不连续排列，下面中部一至二层，二层者内层不连续排列。球果圆柱形或卵状圆柱形，长 7～11 cm，径 3～4 cm，近无梗，成熟前绿色，熟时褐色，中部种鳞肾形，长约 1.5 cm，宽约 2.5 cm，鳞背露出部分密生短毛；苞鳞长约种鳞的 3/4，不外露，上部近圆形，边缘有细缺齿，中央有短急尖头，中下部近等宽，基部渐窄；种子较种翅为长，倒三角状椭圆形，长 8 mm，种翅宽大，倒三角形，上部宽约 1 cm，连同种子长约 1.3 cm。

【花 果 期】 5 月开花，球果 10 月成熟。

【分　　布】 为我国特有树种，产于陕西南部、湖北西部及甘肃南部。

【生　　境】 生长于海拔 2 300～3 000 m 的山沟溪旁及阴地。

【致危因素】 人为砍伐，植株减少，分布区缩减。

[Scientific Name] *Abies chensiensis*

[Morphological Characters] Trees to 50 m tall; branchlets yellowish gray, pale yellow or brownish yellow in 1st year, glabrous or puberulent, branchlets yellowish gray or gray in 2nd and 3rd year; winter buds conical, resinous. Leaves pectinately arranged in 2 lateral sets, linear, 1.5-4.8 cm, dark green adaxially, stomatal lines in 2 white bands abaxially; leaves of cone-bearing branchlets with apex acuminate or obtuse, resin canals 2, marginal or median on cone-bearing branchlets, vegetative branches and young trees have longer leaves, apex 2-cleft or emarginate; a layer of subcutaneous cells from the upper edge to the lower edge of the cross section, which are arranged continuously or discontinuously, one to two layers in the middle of the lower part, and the inner layer is arranged discontinuously. Seed cones cylindric or ovoid-cylindric, 7-11×3-4 cm, subsessile, green, ripening brown, seed scales at middle of cones reniform, ca. 1.5×2.5 cm, exposed part densely pubescent abaxially; bracts ca. 3/4 as long as seed scales, subsessile, distal part orbicular, margin erose-denticulate, apex with short cusp, middle and lower parts nearly equal in width, with narrow base; seeds longer than wing, obtriangular, 8 mm, with broad obtriangular wing, with upper part 1 cm in width, in total 1.3 cm including seed.

[Florescence and Fruit Period] Florescence in May, seed cones mature in October.

[Distribution] Southern Shanxi Province, Western Hubei Province and southern Gansu Province. A native plant of China.

[Habitat] Gullies, stream banks and shady areas with an altitude of 2,300-3,000 m.

[Threatening Factors] Reduction of plant and distribution area caused by felling.

摄像：张金龙

Photo：Zhang Jinlong

34 梵净山冷杉/ Fanjing Shan Lengshan

【学　　名】*Abies fanjingshanensis*

【形态特征】乔木，高达 22 m，胸径 65 cm；树皮暗灰色。一年生枝红褐色，无毛。叶长 1～4.3 cm，宽 2～3 mm，先端凹缺，上面无气孔带，下面气孔带粉白色；树脂道 2，在营养枝上为边生，果枝上则位于叶横切面近两端的叶肉薄壁组织中，近边生。球果圆柱形，熟前紫褐色，熟时深褐色，长 5～6 cm，径约 4 cm，具短柄，中部种鳞肾形，长约 1.5 cm;、宽 1.8～2.2 cm，鳞背露出部分密被短毛；苞鳞长为种鳞的 4/5，上部宽圆，先端微凹或平截，凹处有由中肋延伸的短尖，尖头 1～2 mm，不露出，稀部分露出，种子长卵圆形，微扁，长约 8 mm，种翅褐或灰褐色，连同种子长约 1.5 cm。

【花 果 期】开花期 5—6 月，球果成熟期 9—10 月。

【分　　布】产于贵州。中国特有。

【生　　境】生长于海拔 2 100～2 350 m 的近山脊北山坡林中。

【保护价值】残存种对研究冰川及古植物区系有重要意义。

【致危因素】冰川气候影响，林木生长不良。

【附　　注】缺少影像资料。

[Scientific Name] *Abies fanjingshanensis*

[Morphological Characters] Trees to 22 m tall；trunk to 65 cm d.b.h.；bark dark gray；branchlets red-brown in 1st year，glabrous. Leaves 1-4.3×2-3 cm，apex emarginate，no stomatal bands apically，white-pink stomatal bands abaxially；resin canals 2，marginal on vegetative branchlets，or in the mesophyll parenchyma near the two ends of the leaf cross section on cone-bearing branchlets. Seed cones cylindric，purple-brown，ripening dark brown，5-6×ca. 4 cm，shortly pedunculate，seed scales at middle of cones reniform，ca. 1.5×1.8-2.2 cm，exposed part densely pubescent；bracts spatulate，ca. 4/5 as long as seed scales，distal margin rounded or emarginate，apex slightly concave or flat，with an acute， small cusp extending from the middle ridge, 1-2 mm, cusp not exposed or rarely partly exposed. Seeds slightly appressed, narrowly ovoid, ca. 8 mm, wing brown or gray-brown, in total 1.5 cm with seed.

[Florescence and Fruit Period] Florescence May to June, seed cones mature September to October.

[Distribution] Guizhou Province. A native plant of China.

[Habitat] In the forest on the north mountain slope near the ridge at an altitude of 2,100-2,350 m.

[Protection Value] The remaining species are of great significance to the study of glaciers and palaeoflora.

[Threatening Factors] Under the influence of glacial climate, the growth of trees is not good.

[Note] No image data.

35 资源冷杉/ Ziyuan Lengshan

【学　　名】 *Abies ziyuanensis*

【形态特征】 常绿乔木，高 20～25 m，胸径 40～90 cm；树皮灰白色，片状开裂；1 年生枝淡褐黄色，老枝灰黑色；冬芽圆锥形或锥状卵圆形，有树脂，芽鳞淡褐黄色。叶在小枝上面向外向上伸展或不规则两列，下面的叶呈梳状，线形，长 2～4.8 cm，宽 3～3.5 mm，先端有凹缺，上面深绿色，下面有两条粉白色气孔带，树脂道边生。球果椭圆状圆柱形，长 10～11 cm，直径 4.2～4.5 cm，成熟时暗绿褐色；种鳞扇状四边形，长 2.3～2.5 cm，宽 3～3.3 cm；苞鳞稍较种鳞为短，长 2.1～2.3 cm，中部较窄缩，上部圆形，宽 9～10 mm，先端露出，反曲，有突起的短刺尖；种子倒三角状椭圆形，长约 1 cm，淡褐色，种翅倒三角形，长 2.1～2.3 cm，淡紫黑灰色。树冠高耸于阔叶林层之上，幼树耐阴，大树需要一定的光照，结实有间隔期。

【花 果 期】开花期为 5 月，球果成熟期在 10 月。

【分　　布】间断分布于江西井冈山、湖南、广西资源。中国特有。

【生　　境】生长于海拔 1 400～1 800 m 的高山林中。

【保护价值】为研究古代植物区系有重要意义。

【致危因素】冰川气候影响。难以授粉，种子很难成熟和长成幼树。

[Scientific Name] *Abies ziyuanensis*

[Morphological Characters] Evergreen trees to 20-25 m tall；trunk to 40-90 cm d.b.h.；bark gray，cracking；branchlets pale yellow-brown in 1st year，old branches gray-dark；winter buds conical or conical ovoid，with resin，bud scales pale yellow-brown. Leaves radially spreading or pectinately arranged in 2 lateral sets on upper side of branchlets，unequal，and arranged in comb pattern on lower side，linear，2-4.8 cm×3-3.5 mm，apex emarginate，upper surface dark green，stomatal lines in 2 white-pink bands abaxially，near resin canals. Seed cones oval cylindrical，10-11×4.2-4.5 cm，turning dark green-brown when mature；seed scales fan-quadrilateral，2.3-2.5×3-3.3 cm；bracts broader than seed scales，2.1-2.3 cm in length，middle part narrow，distal margin rounded，9-10 mm in width，with exposed apex，recurved，with short mucro；seeds obtriangular and oval，1 cm in length，pale brown，seed wing obtriangular，2.1-2.3 cm in length，pale lavender-black-gray. The crown of the tree is towering over the broad-leaved forest layer，the young trees are shade tolerant，and the big trees need a certain amount of light. With interval of seed setting.

[Florescence and Fruit Period] Florescence in May，seed cones mature in October.

[Distribution] Intermittently distributed in Jinggangshan，Hunan Province and Guangxi Zhuang Autonomous Region. A native plant of China.

[Habitat] High mountain forest with altitude of 1,400-1,800 m.

[Protection Value] It is of great significance to the study of ancient flora.

[Threatening Factors] Glacial climate impact. It is difficult to pollinate，and seeds are difficult to mature and grow into young trees.

摄像：周建军　徐永福　张金龙

Photo：Zhou Jianjun　Xu Yongfu　Zhang Jinlong

36 银杉/ Yinshan

【学　　名】*Cathaya argyrophylla*

【形态特征】乔木，高达 20 m，胸径 40 cm 以上；树皮暗灰色，老时则裂成不规则的薄片；大枝平展，小枝节间的上端生长缓慢、较粗，或少数侧生小枝因顶芽死亡而成距状，一年生枝黄褐色，密被灰黄色短柔毛，逐渐脱落，二年生枝呈深黄色；叶枕近条形，稍隆起，顶端具近圆形、圆形或近四方状的叶痕，其色较淡；冬芽卵圆形或圆锥状卵圆形，顶端钝，淡黄褐色，无毛，通常长 6～8 mm。叶螺旋状着生成辐射伸展，在枝节间的上端排列紧密，成簇生状，在其之下侧疏散生长，多数长 4～6 cm，宽 2.5～3 mm，边缘微反卷，在横切面上其两端为圆形，下面沿中脉两侧具极显著的粉白色气孔带，每条气孔带有 11～17 行气孔，气孔带一般较浅绿色的叶缘边带为宽；侧生小枝的节间较短，叶排列较密，上端之叶密集、近轮状簇生，多数长不超过 3 cm，边缘平或近平，在横切面上其两端斜尖，下面粉白色气孔带有 9～13 行气孔，宽与边带近相等；叶条形，多少镰状弯曲或直，先端圆，基部渐窄成不明显的叶柄，上面深绿色，被疏柔毛，沿凹陷的中脉有较密的褐色短毛；幼叶上面的毛较多，沿叶缘具睫毛，睫毛不久即脱落，仅留痕迹。雄球花开放前长椭圆状卵圆形，长约 2 cm，径 8～9 mm，盛开时穗状圆柱形，长 5～6 cm，近于无柄，基部围绕的苞片半透明膜质，背面凸起，边缘具不规则的锯齿，位于内部的苞片较大，阔卵形，长 6～8 mm，宽 4～5 mm，外部的苞片多为三角状扁圆形，承托基部的变形叶不久即脱落，雄蕊黄色，长约 6 mm；雌球花基部无苞片，卵圆形或长椭圆状卵圆形，长 8～10 mm，径约 3 mm，珠鳞近圆形或肾状扁圆形，黄绿色，苞鳞黄褐色，三角状扁圆形或三角状卵形，先端具尾状长尖，边缘波状有不规则的细锯齿。球果成熟前绿色，熟时由栗色变暗褐色，卵圆形、长卵圆形或长椭圆形，长 3～5 cm，径 1.5～3 cm，种鳞 13～16 枚，近圆形或带扁圆形至卵状圆形，长 1.5～2.5 cm，宽 1～2.5 cm，背面（尤其是被覆盖着的部分）密被微透明的短柔毛；苞鳞长达种鳞的 1/4～1/3；种子略扁，斜倒卵圆形，基部尖，长 5～6 mm，径 3～4 mm，橄榄绿带墨绿色，有不规则的浅色斑纹，种翅膜质，黄褐色，呈不对称的长椭圆形或椭圆状倒卵形，长 10～15 mm，宽 4～6 mm。

【花 果 期】5 月开花，翌年 10 月球果成熟。

【分　　布】产于湖南，广西金秀、龙胜，重庆南川、武隆，贵州道真、桐梓。中国特有。

【生　　境】生长于海拔 900～1 900 m 的山脊或帽状石山顶部向阳处组成针叶或针阔叶林混生。

【保护价值】木材可供建筑用。为残存于我国的特有单种属植物，其胚胎发育与松属植物相近，具有科研价值。

【致危因素】生态环境破坏，生存能力脆弱。

[Scientific Name] *Cathaya argyrophylla*

[Morphological Characters] Trees to 20 m tall; trunk to 40 cm d.b.h.; bark dark gray, aging irregularly flaking; branches spreading, branchlets joints grow slowly, thick, some lateral branchlets are spur like due to terminal bud death, branchlets yellow-brown in 1st year, densely gray-yellow pubescent, aging glabrous, branchlets dark yellow in 2nd year; Leaf cushions liner, slightly raised, topped with pale, orbicular or subsquare leaf scars; winter buds light yellow-brown, ovoid or ovoid-conical, apex blunt, glabrous, usually 6-8 mm. Leaves radiating outward spirally, arranged on the upper part of branch joints in group, sparse abaxially, often 4-6 cm×2.5-3 mm, margin slightly recurved, two rounded ends on the cross section, with white-pink bands abaxially on both sides of the midvein, 11-17 stomatal lines in each band, with light green margin; lateral branchlets with short joints gap, leaves arranged densely, nearly clustered into a whorl on upper branchlets, usually no longer than 3 cm, margin flat, with two inclined and acuminate ends on the cross section, 9-13 lines of stomatal lines in white-pink bands abaxially, the same with margin band in width; leaves liner, curved or flat, with rounded apex, base contracted into point, upper surface dark green, sparsely tomentose, midvein densely brown tomentose; young leaves densely tomentose, margin with eyelash, glabrescent, with scars remaining. Male cones ovoid or ellipsoid before mature, 2 cm×8-9 mm, and spicate cylinder when mature, 5-6 cm in length, not pedunculate, bracts around base translucent and membranous, raised abaxially, margin pectinate irregularly, large inner bracts, broadly ovate, 6-8×4-5 mm, outer bracts usually triangular-oblate, leaf cushions glabrescent, stamens yellow, 6 mm; female cones has no bracts at base, oval or narrowly oval, 8-10×3 mm, cone scales rounded or reniform, yellow-green, bud scales yellow-brown, triangular oblate or triangular ovate, apex with tail-like mucro, margin wavy and pectinate, irregular. Seed cones green before mature, turning dark brown when mature, oval, oblong or narrowly oval, 3-5×1.5-3 cm, seed scales 13-16, suborbicular or oblate to ovate, 1.5-2.5×1-2.5 cm, translucent puberulent abaxially (specially the covered part); bud scales 1/4-1/3 as long as seed scales; seeds slightly flat, obliquely obovate, base acuminate, 5-6×3-4 mm, olive green with dark green, with irregular light stripes, seed wing membranous, yellow-brown, asymmetrical oval or elliptical obovate, 10-15×4-6 mm.

[Florescence and Fruit Period] Florescence in May， seed cones mature in the next October.

[Distribution] Hunan Province，Jinxiu and Longsheng in Guangxi Zhuang Autonomous Region，Nanchuan and Wulong in Chongqing Municipality，Daozhen and Tongzi in Guizhou Province. A native plant of China.

[Habitat] Coniferous or mixed coniferous broad-leaved forest on the top of mountain ridge or cap stone，900-1,900 m above sea level.

[Protection Value] Wood is available for construction. As a unique single genus plant，its embryonic development is similar to that of Pinus， which has scientific research value.

[Threatening Factors] Destruction of ecological environment and fragile viability.

摄像：林秦文

Photo：Lin Qinwen

37 巧家五针松/ Qiaojia Wuzhensong

【学　　名】*Pinus squamata*

【形态特征】乔木；幼树灰绿色，幼时平滑，老树树皮暗褐色，成不规则薄片剥落，内皮暗白色。冬芽卵球形，红褐色，具树脂。一年生枝红褐色，密被黄褐及灰褐色柔毛，稀有长柔毛及腺体，二年生枝淡绿褐色，无毛。针叶 5（4）针一束，长 9～17 cm，直径约 0.8 mm，两面具气孔线，边缘有细齿，树脂道 3～5，边生，叶鞘早落。成熟球果圆锥状卵圆形，长约 9 cm，直径约 6 cm，果柄长 1.5～2 cm；种鳞长圆状椭圆形，长约 2.7 cm，宽约 1.8 cm，熟时张开，鳞盾显著隆起，鳞脐背生，凹陷，无刺，横脊明显。种子长圆形或倒卵圆形，黑色，种翅长约 1.6 cm，具黑色纵纹。

【花 果 期】花期 4—5 月，球果翌年 9—10 月成熟。

【分　　布】产于云南巧家。中国特有。

【生　　境】生长于海拔 2 200 m 的村旁林中。

【保护价值】科研价值，木材可供家具、建筑用。

【致危因素】物种内在因素；种间影响。

[Scientific Name] *Pinus squamata*

[Morphological Characters] Trees; bark gray-green, aging dark brown, flaking, smooth when young, inner bark pale. Winter buds red-brown, ovoid, resinous. Branchlets red-brown, densely yellow- or gray-brown pubescent in 1st year, rarely long pubescence or glands; branchlets light green in 2nd year, glabrous. Needles 4 or 5 per bundle, 9-17 cm×ca. 0.8 mm, stomatal lines present on all surfaces, margin serrulate, resin canals 3-5, marginal, base with sheath shed. Seed cones pedunculate, 9×6 cm, peduncle 1.5-2 cm; seed scales oblong-elliptic, ca. 2.7×1.8 cm, opened when mature, apophyses swollen, obviously transversely ridged, concave, with no spins, obvious transverse ridge. Seeds black, longitudinally striate, oblong or obovate; wing ca. 1.6 cm, black striate。

[Florescence and Fruit Period] Florescence April to May, seed cones mature in next September to October.

[Distribution] Qiaojia, Yunnan Province. A native plant of China.

[Habitat] In the forest beside a village at an altitude of 2,200 m.

[Protection Value] Scientific research value. Wood can be used for furniture and construction.

[Threatening Factors] Internal factors of species; interspecific effects.

摄像：朱鑫鑫　宋　鼎

Photo：Zhu Xinxin　Song Ding

38 长白松/ Changbaisong

【学　　名】*Pinus sylvestris* var.*sylvestriformis*

【形态特征】乔木，高 20～30 m，胸径 25～40 cm，稀达 1 m；树干通直平滑，基部稍粗糙，棕褐色带黄，龟裂，下中部以上树皮棕黄色至金黄色，裂成鳞状薄片剥落；冬芽卵圆形，芽鳞红褐色，有树脂；一年生枝淡褐色或淡黄褐色，无白粉，二至三年生枝淡灰褐色或灰褐色。针叶 2 枚一束，长 5～8 cm，较粗硬，径 1～1.5 mm；横切面扁半圆形，单层皮下层细胞，维管束之间的距离较宽，树脂道 4～8，边生，稀角部 1～2 个中生或背面 1 个中生。一年生小球果近球形，具短梗，弯曲下垂，种鳞具直伸的短刺；成熟的球果卵状圆锥形，种鳞张开后为椭圆状卵圆形或长卵圆形，长 4～5 cm，径 3～4.5 cm，种鳞背部深紫褐色，鳞盾斜方形或不规则 4～5 角形，灰色或淡褐灰色，强隆起，很少微隆起或近平，球果基部种鳞之鳞盾隆起部分向下弯，横脊明显，纵脊不明显或微明显，鳞脐呈瘤状突起，具易脱落的短刺；种子长卵圆形或三角状卵圆形，长约 4 mm，连翅长约 2 cm，种翅淡褐色，有少数褐色条纹，宽约 7 mm。

【花 果 期】花期 5—6 月上旬，球果翌年 8 月下旬成熟。

【分　　布】产于吉林安图，中国特有。

【生　　境】生长于吉林长白山北坡海拔 800～1 600 m 处，在二道白河以上的林中组成小片纯林；在海拔 1 600 m 的林中则与红松、长白鱼鳞云杉等混生。

【保护价值】木材利用。

【致危因素】林木砍伐。

[Scientific Name] *Pinus sylvestris* var.*sylvestriformis*

[Morphological Characters] Trees to 20-30 m tall; trunk 25-40 cm d.b.h, rarely 1 m, straight, smooth, base slightly rough, yellow-brown, cracked, bark brown to golden above lower middle part, flaking in scale shape; winter buds ovoid, with bud scales red-brown, resinous; branchlets light brown or yellow-brown in 1st year, not glaucous, branchlets light gray or gray-brown in 2nd or 3rd year. Needles 2 per bundle, stiff, 5-8 cm×1-1.5 mm; semiorbicular in cross section, monolayer subcutaneous cell, with large gap between vascular bundles, resin canals 4-8, marginal, rarely 1-2 in the corner or 1 abaxially. Seed cones ovoid in 1st year, with short dropping peduncle, straight short spines on seed scales; seed cones conical-ovoid when mature, turning oval or narrowly oval when the seed scales open, 4-5×3-4.5 cm, seed scales back dark purple-brown, with exposed part diagonal square or irregular 4-5 angle, gray or light brown-gray, apophyses, rarely slightly apophyses or flat, with apophyses part at base downward curved, obviously transversely ridged, longitudinal ridge not or slightly obvious, scale umbilicus with protuberance, and short spines, flaking easily; seeds narrowly ovoid or triangle- ovoid, 4 mm or 2 cm with seed wing, seed wing light brown, with brown striate, 7 mm in width.

[Florescence and Fruit Period] Florescence May to early June, seed cones mature in the next late August.

[Distribution] Antu, Jilin Province. A native plant of China.

[Habitat] On the north slope of Changbai Mountain in Jilin Province, 800-1,600 m above sea level, there are small pure forests in the forests above Erdaobai River. In the forest with an altitude of 1,600 m, the species is mixed with Korean pine, white scale spruce, etc.

[Protection Value] Wood.

[Threatening Factors] Deforestation.

摄像：林秦文

Photo：Lin Qinwen

39 台湾穗花杉/ Taiwan Suihuashan

【学　　名】*Amentotaxus formosana*

【形态特征】小乔木，高达 10 m，胸径 30 cm；大枝稀疏，小枝斜展，微圆或近方形。叶列成两列，披针形或条状披针形，通常微弯镰状，长 5～8.5 cm，宽 5～10 mm，上部渐窄，先端长尖，基部宽楔形或近圆形，几无柄，上面深绿色，中脉微隆起，下面中脉隆起，宽约 1 mm，白色气孔带较绿色边带宽约 1 倍，宽约 2 mm。雄球花穗 2～4 穗，稀 1 穗或 5 穗，长约 3 cm，雄球花几无梗，雄蕊有 5～8 个花药，花丝极短，长不及 1 mm；雌球花近圆球形，有长梗，基部约有 10 枚苞片。种子倒卵状椭圆形或椭圆形，假种皮熟时深红色，长 2～2.5 cm，径 0.9～1.1 cm，顶端有小尖头露出，基部宿存苞片具背脊，梗长 1.5～2 cm。

【花 果 期】花期 4 月，种子翌年 5—6 月成熟。

【分　　布】产于台湾（台东）。中国特有。

【生　　境】生长于海拔 500～1 300 m 的阔叶林中或沟谷中。

【保护价值】木材用于制作家具、工艺品。

【致危因素】生境改变，光照不足。

【附　　注】缺少影像资料。

[Scientific Name] *Amentotaxus formosana*

[Morphological Characters] Trees to 10 m tall；trunk to 30 cm d.b.h.；main branches few. Leafy branchlets broadly ovate-rectangular in outline. Leaves arranged in 2 rows，lanceolate or strip lanceolate，slightly falcate distally，5-8.5 cm×5-10 mm，upper part turning narrow，apex long acuminate，base broadly cuneate or obtuse，subsessile，adaxially dark green，with midvein slightly raised，abaxial midvein raised，1 mm in width，stomatal bands white，ca. 2 mm wide，ca. twice as wide as green marginal bands. Male cone racemes borne racemes borne 2-4，rarely 1 or 5，3 cm in length，subsessile，each with pollen sacs，fine filament, short then 1 mm; female cones ellipsoid, with pedunculate, bracts 10 at base. Seeds obovoid or ellipsoid, finally reddish purple, 2-2.5×0.9-1.1 cm, apex mucronulate, 2 rows of basal scales persistent, pedunculate 1.5-2 cm.

[Florescence and Fruit Period] Florescence in April, seed maturity May to June next year.

[Distribution] East Taiwan. A native plant of China.

[Habitat] In broad-leaved forest or valley with altitude of 500-1,300 m.

[Protection Value] Wood is used to make furniture and crafts.

[Threatening Factors] Habitat change, lack of light.

[Note] No image data.

40 云南穗花杉/ Yunnan Suihuashan

【学　　名】*Amentotaxus yunnanensis*

【形态特征】乔木，高达 15 m，胸径 25 cm；大枝开展，树冠广卵形；小枝向上伸展，微具棱脊，一年生枝绿色或淡绿色，二、三年生枝淡黄色、黄色或淡黄褐色。叶成两列，条形、椭圆状条形或披针状条形，通常直，稀上部微弯，长 3.5～10 cm，稀长达 15 cm，宽 8～15 mm，先端钝或渐尖，基部宽楔形或近圆形，几无柄，边缘微向下反曲，上面绿色，中脉显著隆起，下面淡绿色，中脉近平或微隆起，中脉带宽 1～2 mm，两侧的气孔带干后褐色或淡黄白色，宽 3～4 mm，较绿色边带宽 1 倍或稍宽；萌生枝及幼树之叶的气孔带较窄。雄球花穗 4～6 穗，长 10～15 cm，雄蕊有 4～8（多为 6～7）个花药。种子椭圆形，假种皮成熟时红紫色，微被白粉，长 2.2～3 cm，直径约 1.4 cm，顶端有小尖头露出，基部苞片宿存，背有棱脊，梗较粗，长约 1.5 cm，下部扁平，上部扁四棱形。

【花 果 期】花期 4 月，种子 10 月成熟。

【分　　布】生长于云南东南部海拔约 1 000 m 的石灰岩山地，常组成小片纯林。越南北部也有分布。

【保护价值】木材纹理均匀，结构细致。可供建筑、家具、农具及雕刻等用材。可作庭院树。

【致危因素】长期不当利用，林木采伐，环境改变，森林缩减。

[Scientific Name] *Amentotaxus yunnanensis*

[Morphological Characters] Trees to 15 m tall; trunk to 25 cm d.b.h.; main branches spreading to form a broad, ovoid crown. Leafy branchlets ascending, slightly ridged, axis green or light green in 1st year, thereafter light yellow to yellow to light yellowish brown. Leaves arranged in 2 rows, linear, linear-lanceolate, or linear-elliptic, usually straight, sometimes slightly falcate distally, 3.5-10（15）× 0.8-1.5 cm, apex obtuse or tapered, base broadly cuneate or obtuse, subsessile, margin slightly downcurved, green adaxially, with midvein prominently raised, light green abaxially, with midvein flat or slightly raised, 1-2 mm wide, stomatal bands brown or pale yellowish white when dry, 3-4 mm wide, at least twice as wide as marginal bands; sprouting branches and young tree leaves with narrow stomatal bands. Pollen-cone racemes borne 4-6 together, 10-15 cm, with pollen sacs 4-8（usually 6-7）. Seed ellipsoid, aril reddish ridged, purple when ripe, slightly white powdery, 2.2-3×1.4 cm, apex with short mucro, bracts at base, ridged, thicker pedicel, 1.5 cm, flat abaxially, flattened proximally but quadrangular distally.

[Florescence and Fruit Period] Florescence April, seed mature in October.

[Distribution] Limestone mountain in southeast Yunnan Province, about 1,000 m above sea level, often forms a small pure forest. Also found in northern Vietnam.

[Protection Value] The wood texture is even and the structure is fine. It can be used for building, furniture, farm tools, carving and servicing a garden tree.

[Threatening Factors] Long term improper utilization, tree cutting, environmental change and forest reduction.

摄像：朱鑫鑫　黄江华

Photo：Zhu Xinxin　Huang Jianghua

41 东北红豆杉/ Dongbei Hongdoushan

【学　　名】*Taxus cuspidata*

【形态特征】乔木，高达 20 m，胸径达 1 m；树皮红褐色，有浅裂纹；枝条平展或斜上直立，密生；小枝基部有宿存芽鳞，一年生枝绿色，秋后呈淡红褐色，二、三年生枝呈红褐色或黄褐色；冬芽淡黄褐色，芽鳞先端渐尖，背面有纵脊。叶排成不规则的二列，斜上伸展，约成 45°，条形，通常直，稀微弯，长 1～2.5 cm，宽 2.5～3 mm，稀长达 4 cm，基部窄，有短柄，先端通常凸尖，上面深绿色，有光泽，下面有两条灰绿色气孔带，气孔带较绿色边带宽 2 倍，干后呈淡黄褐色，中脉带上无角质乳头状突起点。雄球花有雄蕊 9～14 枚，各具 5～8 个花药。种子紫红色，有光泽，卵圆形，长约 6 mm，上部具 3～4 钝脊，顶端有小钝尖头，种脐通常三角形或四方形，稀矩圆形。

【花 果 期】花期 5—6 月，种子 9—10 月成熟。

【分　　布】生长于吉林老爷岭、张广才岭及长白山区海拔 500～1 000 m，气候冷湿，酸性土地带，常散生于林中。山东、江苏、江西等省有栽培。

【保护价值】边材窄，黄白色，心材淡褐红色，坚硬、致密，具弹性，有光泽及香气，少反挠，少干裂、比重 0.51。可供建筑、家具、器具、文具、雕刻、箱板等用材；心材可提取红色染料。种子可榨油；木材、枝叶、树根、树皮能提取紫杉素，可治糖尿病；叶有毒，种子的假种皮味甜可食。可作东北及华北地区的庭院树及造林树种。

【致危因素】长期不当利用，林木采伐，环境改变，森林缩减。

[Scientific Name] *Taxus cuspidata*

[Morphological Characters] Trees to 20 m tall; trunk to 1 m d.b.h.; bark reddish brown, with shallow fissures; branchlets spreading or oblique ascending, densely; bud scales persistent at base of branchlets, branchlets green in 1st year, turning pale reddish brown in autumn, reddish brown or yellow-brown in 2nd and 3rd year; winter bud light yellow-brown, tapered apically, ridged dorsally. Leafy branchlets arranged in 2 rows, asymmetric, oblique ascending at 45°, liner, usually straight, rarely curved, 1-2.5 (4) cm×2.5-3 mm, with narrow base, shortly pedunculate, apex usually shortly mucronate, dark green adaxially, glossy, 2 stomatal bands greyish-green, at least twice as wide as green marginal bands, tawny yellow when dry, midvein not papillate abaxially. Microsporophylls 9-14, each with 5-8 pollen sacs. Seed purplish red, glossy, ovoid, 6 mm, distally with 3 or 4 or more obtuse ridges, apex with small, obtuse mucro, hilum usually triangular or quadrangular, rarely oblong.

[Florescence and Fruit Period] Florescence May to June, seed maturity September to October.

[Distribution] Laoye Ridge, Zhangguangcai Ridge and Mount Changbai in Jilin Province, 500-1,000 m above sea level. Cold and humid climate, acid soil zone, often scattered in the forest. Cultivated in Shandong Province, Jiangsu Province, Jiangxi Province and other provinces.

[Protection Value] The sapwood is narrow, yellow and white; the heartwood is light brown red, hard, dense, elastic, glossy and fragrant, with less dry crack and specific gravity of 0.51. The wood used for building, furniture, appliance, stationery, carving, box board, etc.; the heartwood yields a red dye, oil is extracted from the seeds, and taxol used to treat diabetes is extracted from the wood, bark, leaves, and roots; the leaves are poisonous and the aril of seeds is sweet and edible. It can be used as garden tree and afforestation tree species in Northeast and North China.

[Threatening Factors] Long term improper use, tree cutting, environmental change, forest reduction.

摄像：林秦文

Photo: Lin Qinwen

42 喜马拉雅红豆杉/ Ximalaya Hongdoushan

【学　　名】 *Taxus wallichiana*

【形态特征】 乔木或大灌木；一年生枝绿色，干后呈淡褐黄色、金黄色或淡褐色，二、三年生枝淡褐色或红褐色；冬芽卵圆形，基部芽鳞的背部具脊，先端急尖。叶条形，较密地排列成彼此重叠的不规则两列，质地较厚，通常直，上下几等宽或上端微渐窄，先端有突起的刺状尖头，基部两侧对称，边缘不反曲或反曲（干叶明显），上面光绿色，下面沿中脉带两侧各有一条淡黄色气孔带，中脉带与气孔带上均密生均匀细小角质乳头状突起点。种子生于红色肉质杯状的假种皮中，柱状矩圆形，上下等宽或上部较宽，微扁，长约 6.5 mm，直径 4.5～5 mm，上部两侧微有钝脊，顶端有凸起的钝尖，种脐椭圆形。

【花 果 期】 花期 4—5 月，种子 9—10 月成熟。

【分　　布】 四川、云南、西藏，尼泊尔、不丹、印度、巴基斯坦有分布。

【生　　境】 分布于海拔 2 500～3 000 m 地带针阔混交。

【保护价值】 木材心边材区别明显，纹理均匀，结构细致，硬度大，韧性强，干后少挠裂。为优良的建筑、桥梁、家具、器具、车辆等用材。可作产区的造林树种。

【致危因素】长期不当利用，林木采伐，环境改变，森林缩减。

[Scientific Name] *Taxus wallichiana*

[Morphological Characters] Trees or shrubs；branchlets green in 1st year，turning light brown yellow，golden yellow or light brown when dry，light brown or reddish brown in 2nd or 3rd year；winter bud ovoid，bud scales at base，with ridges abaxially，apex acuminate. Leaves liner，arranged in 2 rows，irregular，densely，overlapping，thick，usually straight，almost equally wide throughout length，or distal，or distal part slightly narrow，apex acuminate，base symmetric，margin recurved or not recurved，（leaves obvious），green adaxially，2 stomatal bands along midvein abaxially，pale yellowish，densely and evenly papillate on midvein bands and stomatal bands. Seeds in red，fleshy and cup-shaped aril，columnar-rectangle，equally wide throughout length，or distal part wider，slightly flattened，6.5×4.5-5 mm，with obtuse ridges adaxially，apex with small mucro，hilum elliptic.

[Florescence and Fruit Period] Florescence April to May，seed maturity September to October.

[Distribution] Sichuan Province，Yunnan Province，Tibet；Nepal，Bhutan，India and Pakistan.

[Habitat] Coniferous and broad-leaved mixed forest in the area of 2,500-3,000 m above sea level.

[Protection Value] The difference between heartwood and sapwood is obvious，the texture is even，the structure is fine，the hardness is large，the toughness is strong，and there is less crack after drying. It is used for excellent buildings，bridges，furniture，appliances，vehicles and other materials. Tree species for afforestation in production area.

[Threatening Factors] Long term improper use，tree cutting，environmental change，forest reduction.

摄像：林秦文

Photo：Lin Qinwen

43 红豆杉/ Hongdoushan

【学　　名】*Taxus wallichiana* var. *chinensis*（Taxus chinensis）

【形态特征】乔木，高达 30 m，胸径达 60～100 cm；树皮灰褐色、红褐色或暗褐色，裂成条片脱落；大枝开展，一年生枝绿色或淡黄绿色，秋季变成绿黄色或淡红褐色，二、三年生枝黄褐色、淡红褐色或灰褐色；冬芽黄褐色、淡褐色或红褐色，有光泽，芽鳞三角状卵形，背部无脊或有纵脊，脱落或少数宿存于小枝的基部。叶排列成两列，条形，微弯或较直，长 1～3（多为 1.5～2.2）cm，宽 2～4（多为 3）mm，上部微渐窄，先端常微急尖，稀急尖或渐尖，上面深绿色，有光泽，下面淡黄绿色，有两条气孔带，中脉带上有密生均匀而微小的圆形角质乳头状突起点，常与气孔带同色，稀色较浅。雄球花淡黄色，雄蕊 8～14 枚，花药 4～8（多为 5～6）枚。种子生于杯状红色肉质的假种皮中，间或生于近膜质盘状的种托（即未发育成肉质假种皮的珠托）之上，常呈卵圆形，上部渐窄，稀倒卵状，长 5～7 mm，径 3.5～5 mm，微扁或圆，上部常具二钝棱脊，稀上部三角状具三条钝脊，先端有突起的短钝尖头，种脐近圆形或宽椭圆形，稀三角状圆形。

【花 果 期】花期 4—5 月，种子 9—10 月成熟。

【分　　布】产于浙江、安徽、河南、湖北、湖南、广西、重庆、四川等。中国特有。

【生　　境】生长于海拔 1 000～1 200 m 以上山地。

【保护价值】为优良的建筑、桥梁、家具、器具、车辆等用材，其树皮、枝、叶、种子可提取抗癌药物紫杉醇。

【致危因素】长期不当利用，林木采伐，环境改变，森林缩减。

[Scientific Name] *Taxus wallichiana* var. *chinensis*（Taxus chinensis）

[Morphological Characters] Trees to 30 m tall; trunk to 60-100 cm d.b.h.; bark grayish brownish，reddish or dark brown，peeling off in strips or cracking and falling off; main branches spreading，branches green or light yellow-green in 1st year，turning green-yellow or light reddish brown in autumn，yellow-brown，light reddish brown or grayish brown in 2nd and 3rd year; winter bud yellow-brown，light brown or reddish brown，glossy，bud scales triangular-ovate，with or without longitudinal ridges abaxially，falling off or at the branchlets base. Leaves arranged in 2 rows，liner，slightly curved or straight，1-3（usually 1.5-2.2）cm×2-4（usually3）mm，apex gradually acuminate or abruptly tapered and indistinctly mucronate，dark green adaxially，light yellow-green abaxially，with stomatal bands 2，densely and evenly papillate on midvein band，usually in the same color with stomatal band，rarely in light color. Male cones light yellow，stamens 8-14，pollen sacs 4-8（usually 5-6）. Seeds in red，fleshy and cup-shaped aril，or on membranous and disciform seed tray（undeveloped fleshy aril），usually ovate，distal attenuate，rarely obovoid，5-7×3.5-5 mm，slightly flattened，usually with 2 obtuse ridges，rarely with triangle 3 obtuse ridges，apex with short obtuse mucro，hilum elliptic or broad elliptic，rarely rounded-trigonous.

[Florescence and Fruit Period] Florescence April to May，seed maturity September to October.

[Distribution] Zhejiang Province，Anhui Province，Henan Province，Hubei Province，Hunan Province，Guangxi Zhuang Autonomous Region，Chongqing Municipality，Sichuan Province，etc. A native plant of China.

[Habitat] Mountains over 1,000-1,200 m above sea level.

[Protection Value] It is a good building，bridge，furniture，appliance，vehicle and other materials. Its bark，branches，leaves and seeds can extract taxol，an anticancer drug.

[Threatening Factors] Long term improper use，tree cutting，environmental change，forest reduction.

摄像：林秦文

Photo：Lin Qinwen

44 南方红豆杉/ Nanfang Hongdoushan

【学　　名】*Taxus wallichiana* var. *mairei*

【形态特征】乔木，高达 30 m，胸径达 60～100 cm；树皮灰褐色、红褐色或暗褐色，裂成条片脱落；大枝开展，一年生枝绿色或淡黄绿色，秋季变成绿黄色或淡红褐色，二、三年生枝黄褐色、淡红褐色或灰褐色；冬芽黄褐色、淡褐色或红褐色，有光泽，芽鳞三角状卵形，背部无脊或有纵脊，脱落或少数宿存于小枝的基部。叶排列成两列，条形，微弯或较直，长 1～3（多为 1.5～2.2）cm，宽 2～4（多为 3）mm，上部微渐窄，先端常微急尖，稀急尖或渐尖，上面深绿色，有光泽，下面淡黄绿色，有两条气孔带，中脉带上有密生均匀而微小的圆形角质乳头状突起点，常与气孔带同色，稀色较浅。雄球花淡黄色，雄蕊 8～14 枚，花药 4～8（多为 5～6）枚。种子生于杯状红色肉质的假种皮中，间或生于近膜质盘状的种托（即未发育成肉质假种皮的珠托）之上，常呈卵圆形，上部渐窄，稀倒卵状，长 5～7 mm，径 3.5～5 mm，微扁或圆，上部常具二钝棱脊，稀上部三角状具三条钝脊，先端有突起的短钝尖头，种脐近圆形或宽椭圆形，稀三角状圆形。

【花 果 期】花期 4—5 月，种子 9—10 月成熟。

【分　　布】产于甘肃南部、陕西南部、四川、云南东北部及东南部、贵州西部及东南部、湖北西部、湖南东北部、广西北部和安徽南部（黄山）。

【生　　境】生长于海拔 1 000～1 200 m 以上的高山上部。

【保护价值】心材橘红色，边材淡黄褐色，纹理直，结构细，坚实耐用，干后少开裂。可供建筑、车辆、家具、器具、农具及文具等用材。其树皮、枝、叶、种子可提取抗癌药物紫杉醇。

【致危因素】长期不当利用，林木采伐，环境改变，森林缩减。

[Scientific Name] *Taxus wallichiana* var. *mairei*

[Morphological Characters] Trees to 30 m tall; trunk to 60-100 cm d.b.h.; bark grayish brownish, reddish or dark brown, peeling off in strips or cracking and falling off; main branches spreading, branches green or light yellow-green in 1st year, turning green-yellow or light reddish brown in autumn, yellow-brown, light reddish brown or grayish brown in 2nd and 3rd year; winter bud yellow-brown, light brown or reddish brown, glossy, bud scales triangular-ovate, with or without longitudinal ridges abaxially, falling off or at the branchlets base. Leaves arranged in 2 rows, liner, slightly curved or straight, 1-3 (usually 1.5-2.2) cm× 2-4(usually 3)mm, apex gradually acuminate or abruptly tapered and indistinctly mucronate, dark green adaxially, glossy, light yellow-green abaxially, with stomatal bands 2, densely and evenly papillate on midvein band, usually in the same color with stomatal band, rarely in light color. Male cones light yellow， stamens 8-14， pollen sacs 4-8（usually 5-6）. Seeds in red， fleshy and cup-shaped aril， or on membranous and disciform seed tray（undeveloped fleshy aril）, usually ovate，distal attenuate，rarely obovoid，5-7×3.5-5 mm，slightly flattened or round， usually with 2 obtuse ridges， rarely with triangle 3 obtuse ridges， apex with short obtuse mucro， hilum elliptic or broad elliptic， rarely rounded-trigonous.

[Florescence and Fruit Period] Florescence April to May， seed maturity September to October.

[Distribution] South Gansu Province， South Shaanxi Province， Sichuan Province， Northeast and Southeast Yunnan Province，West and Southeast Guizhou Province，West Hubei Province，Northeast Hunan Province，North Guangxi Zhuang Autonomous Region and South Anhui Province（Mount Huang）.

[Habitat] The upper part of the mountain with an altitude of 1,000-1,200 m.

[Protection Value] The heartwood is orange red，the sapwood is light yellow brown，with straight texture，fine structure，solid and durable，and less cracking after drying. It can be used for construction， vehicles， furniture， appliances， farm tools， stationery and other materials. Taxol can be extracted from bark， branches， leaves and seeds.

[Threatening Factors] Long term improper use， tree cutting， environmental change， forest reduction.

摄像：林秦文

Photo：Lin Qinwen

45 水松/ Shuisong

【学　　名】*Glyptostrobus pensilis*

【形态特征】乔木，高 8～10 m，稀高达 25 m，生长于湿生环境，树干基部膨大成柱槽状，并且有伸出土面或水面的吸收根，柱槽高达 70 cm，干基直径达 60～120 cm，树干有扭纹；树皮褐色或灰白色而带褐色，纵裂成不规则的长条片；枝条稀疏，大枝近平展，上部枝条斜伸；短枝从二年生枝的顶芽或多年生枝的腋芽伸出，长 3～18 cm，冬季脱落；主枝则从多年生及二年生的顶芽伸出，冬季不脱落。叶多型：鳞形叶较厚或背腹隆起，螺旋状着生于多年生或当年生的主枝上，长约 2 mm，有白色气孔点，冬季不脱落；条形叶两侧扁平，薄，常列成二列，先端尖，基部渐窄，长 1～3 cm，宽 1.5～4 mm，淡绿色，背面中脉两侧有气孔带；条状钻形叶两侧扁，背腹窿起，先端渐尖或尖钝，微向外弯，长 4～11 mm，辐射伸展或列成三列状；条形叶及条状钻形叶均于冬季连同侧生短枝一同脱落。球果倒卵圆形，长 2～2.5 cm，径 1.3～1.5 cm；种鳞木质，扁平，中部的倒卵形，基部楔形，先端圆，鳞背近边缘处有6～10个微向外反的三角状尖齿；苞鳞与种鳞几全部合生，仅先端分离，三角状，向外反曲，位于种鳞背面的中部或中上部；种子椭圆形，稍扁，褐色，长5～7 mm，宽3～4 mm，下端有长翅，翅长4～7 mm。子叶4～5枚，条状针形，长1.2～1.6 cm，宽不及1 mm，无气孔线；初生叶条形，长约2 cm，宽1.5 mm，轮生、对生或互生，主茎有白色小点。

【花 果 期】花期1—2月，球果秋后成熟。

【分　　布】为我国特有树种，主要分布在广州珠江三角洲和福建中部及闽江下游海拔1 000 m 以下地区。广东东部及西部、福建西部及北部、江西东部、四川东南部、广西及云南东南部也有零星分布。此外南京、武汉、庐山、上海、杭州等地有栽培。

【生　　境】生长于河流两岸。

【保护价值】木材淡红黄色，材质轻软，纹理细，耐水湿，可作建筑、桥梁、家具等用材。根部的木质轻松，浮力大，可做救生圈、瓶塞等软木用具。种鳞、树皮含单宁，可染渔网或制皮革。树形优美，可作庭院树种。

【致危因素】过度采伐利用。

[Scientific Name] *Glyptostrobus pensilis*

[Morphological Characters] Trees to 8-10 m, rarely 25 m, in moist environment, dilated at base with groove, absorption roots extending out of the soil surface or water surface, groove 70 cm, trunk base 60-120 cm in diameter, with twisted lines; bark brown or grayish white with brown tinge, cracking into long, irregular strips; branches spare, main branches spreading horizontally, with distal branchlets obliquely stretching; branchlets extending from the terminal bud of a biennial branch or the axillary bud of a perennial branch, 8-18 cm, falling off in winter; main branches extending from the terminal buds of perennial and biennial, not falling in winter. Leaves multiple type: scale-like leaves thick or raised abaxially, spiraling on a perennial or annual main branch, 2 mm in length, with scattered, white, stomatal spots, not falling in winter; liner leaves flattened, thin, usually in 2 rows, apex acuminate, base tapered, 1-3 cm×1.5-4 mm, light green, with stomatal bands along midvein abaxially; liner subulate leaves, flattened on both sides, both surfaces raised, apex gradually acuminate or abruptly tapered and indistinctly mucronate, slightly outwards curved, 4-11 mm, extending in 3 rows; liner leaves and liner- subulate leaves fall in winter with lateral branchlets. Seed cones obovate, 2-2.5×1.3-1.5 cm; cone scales woody, flattened, median scales obovate, base cuneate, apex rounded, apical margin with 6-10 triangular, outwardly curved teeth adaxially; bud scales and seed scales all united, only apex separated, triangle, outwards curved, on middle or upper part of abaxial seed scales; Seeds elliptic, slightly flattened, brown, 5-7×3-4 mm, with a basal wing 4-7 mm. Cotyledon 4-5, liner-lanceolate, 1.2-1.6 cm×1 mm, with no stomatal lines; primary leaves liner, 2 cm×1.5 mm, whorled, opposite, or alternate, with white spots on main stem.

[Florescence and Fruit Period] Florescence January to February, seed cones mature in autumn.

[Distribution] A native plant of China. It is mainly distributed in the Pearl River Delta of Guangzhou, the central part of Fujian Province and the lower reaches of Minjiang River with an altitude of less than 1,000 m. There is also sporadic distribution in the East and West of Guangdong Province, the West and North of Fujian Province, the East of Jiangxi Province, the Southeast of Sichuan Province, Guangxi Zhuang Autonomous Region and the Southeast of Yunnan Province. In addition，it is cultivated in Nanjing，Wuhan，Lushan，Shanghai，Hangzhou and other places.

[Habitat] River banks.

[Protection Value] The wood is light red and yellow，light and soft，with fine texture and moisture resistance. It can be used as building，bridge，furniture and other materials. The wood at the root is easy and buoyant. It can be used as life buoy，cork and other cork tools. The seed scales and bark contain tannins，which can be dyed into fishnets or leather. It has a beautiful shape and can be used as garden tree.

[Threatening Factors] Over harvesting.

摄像：徐晔春　徐永福　朱鑫鑫

Photo：Xu Yechun　Xu Yongfu　Zhu Xinxin

46 水杉/ Shuishan

【学　　名】*Metasequoia glyptostroboides*

【形态特征】乔木，高达 35 m，胸径达 2.5 m；树干基部常膨大；树皮灰色、灰褐色或暗灰色，幼树裂成薄片脱落，大树裂成长条状脱落，内皮淡紫褐色；枝斜展，小枝下垂，幼树树冠尖塔形，老树树冠广圆形，枝叶稀疏；一年生枝光滑无毛，幼时绿色，后渐变成淡褐色，二、三年生枝淡褐灰色或褐灰色；侧生小枝排成羽状，长 4～15 cm，冬季凋落；主枝上的冬芽卵圆形或椭圆形，顶端钝，长约 4 mm，径 3 mm，芽鳞宽卵形，先端圆或钝，长宽几乎相等，2～2.5 mm，边缘薄而色浅，背面有纵脊。叶条形，长 0.8～3.5（常 1.3～2）cm，宽 1～2.5（常 1.5～2）mm，上面淡绿色，下面色较淡，沿中脉有两条较边带稍宽的淡黄色气孔带，每带有 4～8 条气孔线，叶在侧生小枝上列成二列，羽状，冬季与枝一同脱落。球果下垂，近四棱状球形或矩圆状球形，成熟前绿色，熟时深褐色，长 1.8～2.5 cm，直径 1.6～2.5 cm，梗长 2～4 cm，其上有交对生的条形叶；种鳞木质，盾形，通常 11～12 对，交叉对生，鳞顶扁菱形，中央有一条横槽，基部楔形，高 7～9 mm，能育种鳞有 5～9 粒种子；种子扁平，倒卵形，间或圆形或矩圆形，周围有翅，先端有凹缺，长约 5 mm，径 4 mm；子叶 2 枚，条形，长 1.1～1.3 cm，宽 1.5～2 mm，两面中脉微隆起，上面有气孔线，下面无气孔线；初生叶条形，交叉对生，长 1～1.8 cm，下面有气孔线。

【花 果 期】花期 2 月下旬，球果 11 月成熟。

【分　　布】仅分布于四川石柱县及湖北利川县磨刀溪、水杉坝一带及湖南西北部龙山及桑植等地。

【生　　境】生长于海拔 750～1 500 m、气候温和、夏秋多雨、酸性黄壤土地区。在河流两旁、湿润山坡及沟谷中栽培很多，也有少数野生树木，常与杉木、茅栗、锥栗、枫香、漆树、灯台树、响叶杨、利川润楠等树种混生。

【保护价值】边材白色，心材褐红色，材质轻软，纹理直，结构稍粗，早晚材硬度区别大，不耐水湿。可供房屋建筑、板料、电杆、家具及木纤维工业原料等用。

【致危因素】冰川气候影响以及人为采伐。

[Scientific Name] *Metasequoia glyptostroboides*

[Morphological Characters] Trees to 35 m tall; trunk buttressed at base, to 2.5 m d.b.h.; bark gray, grayish brown or dark brown, with thin flakes and exfoliating when young, and strips and exfoliating when mature, inner bark light purplish brown; branches oblique ascending, with branchlets pendulous, crown pyramidal when young, finally broadly conical, with sparse leaves; axis smooth in 1st year, green when young, growing to be light brown, or light grayish brown to gray-brown in 2nd and 3rd year; lateral branchlets deciduous, 4-15 cm, falling in winter; winter buds on main branch ovoid or oblong-ellipsoid, apex obtuse, 4×3 mm, bud scales broadly ovoid, apex obtuse or rounded, 2-2.5×2-2.5 mm, margin thin in light color, with longitudinal ridges abaxially. Leaves liner, 0.8-3.5 (usually 1.3-2) cm×1-2.5 (usually 1.5-2) cm, light green adaxially, light colored abaxially, stomatal bands 2 pale yellowish, along midvein, broader than margin bands, with 4-8 stomatal lines each, leaves arranged in 2 rows on lateral branchlets, pinnate, falling off in winter with branchlets. Seed cones pendulous, quadrangular sphere or rectangular sphere, green before mature, dark brown when mature, 1.8-2.5×1.6-2.5 cm, pedunculate 2-4 cm, with opposite liner leaves; seed scales woody, peltate, usually 11-12 pairs, crossing opposite, apex flat subulate, with a groove in the middle, base cuneate, 7-9 mm in length, fertile scale with 5-9 seeds; seeds flat, obovate, or oblong-ellipsoid, with a basal wing, apex concave, 5-4 mm; cotyledon 2, liner, 1.1-1.3 cm×1.5-2 mm, midvein slightly raised on both surfaces, with stomatal lines adaxially, no stomatal lines abaxially; primary leaves liner, crossing, 1-1.8 cm in length, with stomatal lines abaxially.

[Florescence and Fruit Period] Florescence in late February, seed cones in November.

[Distribution] It is only distributed in Shizhu County of Sichuan Province, River Modao and Shuishanba of Lichuan County of Hubei Province, Longshan and Sangzhi in the Northwest of Hunan Province.

[Habitat] 750-1,500 m above sea level, mild climate, rainy summer and autumn, acid yellow soil area. There are many wild trees on both sides of the river, on the wet hillside and in the valley, often mixed with Abies holophylla, Castanea seguinii, Castanea henryi,

Liquidambar， lacquer tree， lampstand tree， ring leaf poplar， Lichuan runnan， etc.

[Protection Value] Sapwood is white， heartwood is maroon， material is light and soft， texture is straight， structure is slightly thick， hardness of early and late wood is different greatly， and it is not water resistant. It can be used for building， board， pole， furniture and wood fiber industrial raw materials.

[Threatening Factors] Glacial climate impact and man-made logging.

摄像：林秦文

Photo：Lin Qinwen

47 长喙毛茛泽泻/ Changhui Maogenzexie

【学　　名】*Ranalisma rostratum*

【形态特征】多年生沼生或水生草本。根状茎匍匐。叶多数，基生，叶片薄纸质，全缘；沉水叶披针形，长 3～7 cm，宽 1～1.5 cm，先端渐尖；浮水叶或挺水叶卵圆形、卵状椭圆形，长 3～4.5 cm，宽 3～3.5 cm，先端钝尖，基部浅心形；叶柄长 12～22 cm。花葶直立，高约 20 cm 或更高。花 1～3 朵，着生于花葶顶部，苞片 2 枚，佛焰苞状，长约 7 mm。花两性；外轮花被片 3 枚，绿色，广椭圆形，长约 5 mm，先端钝圆；内轮花被片与外轮近等长，倒卵状椭圆形；心皮多数，密集于花托上，花柱顶生，长于心皮，宿存；雄蕊 9 枚，长约 2.5 mm，花药椭圆形，花丝上下等宽；花托突起，呈柱状。果实多少侧扁，近倒三角形，顶端具喙，长 3～5 mm，基部宽 1.5～3 mm，向上渐尖呈芒状。

【花 果 期】花果期 8—9 月。

【分　　布】产于浙江丽水四明山、江西和湖南。越南、印度、马来西亚及非洲热带地区有分布。

【生　　境】生长于池沼浅水中。

【保护价值】观赏、科研。

【致危因素】生境破坏与丧失，自然种群过小。

[Scientific Name] *Ranalisma rostratum*

[Morphological Characters] Herb, perennial marsh or aquatic. Rhizome stolon. Leaves basal, leaf blade papery, with complete margin; submerged leaf lanceolate, 3-7×1-1.5 cm, lanceolate; floating or emergent leaves elliptic or ovate, 3-4.5×3-3.5 cm, apex with obtuse mucro, base heart-shaped; petiole 12-22 cm. Scape straight, 20 cm or taller. Flowers 1-3 whorled at apex of scape, with bracteate 2, in spathe shape, 7 mm. Flower bisexual; with outer bracteate 3, green, broadly elliptic, 5 mm, apex obtuse; inner and outer bracteate with same length, obovate; carpels numerous, on scape, style persistent, longer than carpel; Stamens 9, 2.5 mm, anthers ellipsoid, filaments upper and lower equal width; receptacle raised, columnar, fruit flattened, obtuse triangle, with margin at apex 3-5 mm, base 1.5-3 mm in width, upwards tapered.

[Florescence and Fruit Period] Florescence and fruit period August to September.

[Distribution] Siming Mountain, Lishui, Zhejiang Province, Jiangxi Province and Hunan Province. Also found in Vietnam, India, Malaysia and tropical Africa.

[Habitat] In shallow water.

[Protection Value] Ornamental plant and scientific research.

[Threatening Factors] Habitat destruction and loss, natural population is too small.

摄像：刘　军

Photo: Liu Jun

48 普陀鹅耳枥/ Putuo E'erli

【学　　名】*Carpinus putoensis*

【形态特征】乔木；树皮灰色；小枝棕色，疏被长柔毛和黄色椭圆形小皮孔，后渐无毛而呈灰色。叶厚纸质，椭圆形至宽椭圆形，长 5～10 cm，宽 3～5 cm，顶端锐尖或渐尖，基部圆形或宽楔形，边缘具不规则的刺毛状重锯齿，上面疏被长柔毛，下面疏被短柔毛，以后两面均渐变无毛，仅下面沿脉密被短柔毛及脉腋间具簇生的髯毛，侧脉 11～13 对；叶柄长 5～10 mm，上面疏被短柔毛。果序长 3～8 cm，直径 4～5 cm；序梗、序轴均疏被长柔毛或近无毛、序梗长 1.5～3 cm；果苞半宽卵形，长约 3 cm，背面沿脉被短柔毛，内侧基部具长约 3 mm 的内折的卵形小裂片，外侧基部无裂片，中裂片半宽卵形，长约 2.5 cm，顶端圆或钝，外侧边缘具不规则的齿牙状疏锯齿，内侧边缘全缘，直或微呈镰形。小坚果宽卵圆形，长约 6 mm，无毛亦无腺体，具数肋。

【花 果 期】花期 6—7 月，果期 8—9 月。

【分　　布】我国特有种，仅见于浙江舟山群岛。

【生　　境】生长于山坡林中。

【保护价值】数量稀少，珍贵，特有。

【致危因素】物种内在因素；生境退化或丧失。

[Scientific Name] *Carpinus putoensis*

[Morphological Characters] Tree; bark brown-gray, branchlets brown, sparsely villous, with yellow oval pinhole, turning glabrous and gray. Leaf blade papery, elliptic or broadly elliptic, 5-10×3-5 cm, apex acuminate or tapered, base rounded or broadly cuneate, margin irregularly and doubly setiform serrate, adaxially sparsely villous, abaxially sparsely pubescent, glabrescent, with pubescent along veins and bearded in axils of lateral veins remain, lateral veins 11-13 pairs; petiole 5-10 mm, sparsely pubescent; female inflorescence, 3-8 cm in length, 4-5 cm in diameter; peduncle 1.5-3 cm, sparsely villous or glabrescent; bracts semiovate, 3 cm, abaxially pubescent along veins, with ovate, inflexed basal lobe, 3 mm, without basal lobe, middle lobe semiovate 2.5 cm, apex rounded or obtuse, outer margin irregularly remotely serrate, inner margin entire or subundulate. Nutlet broadly ovoid, ca. 6 mm, glabrous, with no glands, prominently ribbed.

[Florescence and Fruit Period] Florescence June to July, fruit August to September.

[Distribution] A native plant of China, only found in Zhoushan Islands, Zhejiang Province.

[Habitat] Hillside forest.

[Protection Value] Rare in quantity; precious; peculiar.

[Threatening Factors] Internal factors of species; habitat degradation or loss.

摄像：刘 军 武 晶

Photo: Liu Jun Wu Jing

49 天目铁木/ Tianmu Tiemu

【学　　名】*Ostrys rehderiana*

【形态特征】乔木，高达 18 m，胸径达 45 cm；树皮深灰色，粗糙；枝条灰褐色或暗灰色，无毛，皮孔疏生；小枝细瘦，褐色，幼时密被短柔毛，以后渐变疏至几无毛，具条棱，疏生皮孔；芽长卵圆形，长约 5 mm，锐尖，芽鳞亮绿色，覆瓦状排列，卵形，渐尖，无毛。叶长椭圆形或矩圆状卵形，长 3～10 cm，宽 1.8～4 cm；顶端渐尖、长渐尖或尾状渐尖，基部近圆形或宽楔形；边缘具不规则的锐齿或有时具刺毛状齿；叶上面绿色，几无毛，下面淡绿色，疏被硬毛至几无毛；叶脉在上面微陷，沿中脉密被短柔毛，在下面隆起，疏被短硬毛间或有短柔毛，脉腋间有时具髯毛，侧脉 13～16 对，脉间相距 4～7 mm；叶柄长 3～5 mm，密被短柔毛。雄花序下垂，长 5～10 cm，单生或 2～3 枚簇生；苞鳞宽卵形，顶端骤尖，具条棱，边缘密生短纤毛；花药顶端具长柔毛。果多数，聚生成稀疏的总状；果序轴全长 2～3 cm，序梗长 1.5～2 cm，密被短硬毛；果苞膜质，膨胀，长椭圆形至倒卵状披针形，长 2～2.5 cm，最宽处直径 6～8 mm，顶端圆，具短尖，基部缢缩呈柄状，上部无毛，基部具长硬毛，网脉显著。小坚果红褐色，卵状披针状，长 7～8 mm，直径约 2.5 mm，平滑，具不明显的细肋。

【花 果 期】花期 6—7 月，果期 8—9 月。

【分　　布】浙江天目山。中国特有种。

【生　　境】生长于海拔 400～500 m 的林中。

【保护价值】天目铁木不仅是我国特有种，而且是该属分布于我国东部的唯一种类。对研究植物区系和铁木属系统分类，以及保存物种等，均具有一定意义。仅存 6 株。
【致危因素】生境退化或丧失。

[Scientific Name] *Ostrya rehderiana*
[Morphological Characters] Trees to 18 m tall; trunk to 45 cm d.b.h; bark gray, rough; branches brown-gray or dark gray, glabrous, lenticels sparsely; branchlets brown, slender, sericeous-pubescent when young, becoming glabrous, with line ridges, lenticels sparsely; buds narrowly oblong, 5 mm in length, apex acuminate, bud scales bright green, imbricate, oblong, acuminate, glabrous. Leaf blade elliptic-oblong or narrowly oblong, 3-10×1.8-4 cm; apex acuminate or caudate-acuminate, base subrounded or broadly cuneate; margin irregularly and doubly sharply serrate, sometimes setiform serrate; leaves green adaxially, glabrous, light green abaxially, sparsely hirtellous or glabrous; veins slightly sags adaxially, densely pubescent along midvein, abaxially raised, sparsely pubescent and hirtellous, bearded in axils of lateral veins, lateral veins 13-16 pairs, gap 4-7 mm; petiole 3-5 mm, densely pubescent. Male inflorescence pendulous, 5-10 cm in length, single or 2-3 in groups; bracts obovate-elliptic, apex acuminate, obscurely ribbed, margin densely short pubescent; anther apex villous. Fruit numerous, sparsely racemes; fruit axis 2-3 cm, stipe 1.5-2 cm, densely short hirtellous; bracts membranous, expended, obovate-elliptic or obovate-lanceolate, 2-2.5 cm×6-8 mm, apex rounded, apiculate, base constricted into a stipe, distally glabrous, basally long hirtellous, reticulate veins obvious. Nutlet red-brown, narrowly oblong,

7-8×2.5 mm，glabrous，obscurely ribbed.

[Florescence and Fruit Period] Florescence June to July，fruit August to September.

[Distribution] Mount Tianmu in Zhejiang Province. A native plant in China.

[Habitat] In the forest 400-500 m above sea level.

[Protection Value] *Ostrya rehderiana* is not only an endemic species in China，but also the only species of this genus distributed in the east of China. It is of great significance to the study of flora，taxonomy of Hemerocallis and conservation species. There are only six left.

[Threatening Factors] Habitat degradation or loss.

摄像：汪　远

Photo：Wang Yuan

50 伯乐树/ Boleshu

【学　　名】 *Bretschneidera sinensis*

【形态特征】 乔木，高 10～20 m；树皮灰褐色；小枝有较明显的皮孔。羽状复叶通常长 25～45 cm，总轴有疏短柔毛或无毛；叶柄长 10～18 cm，小叶 7～15 片，纸质或革质，狭椭圆形，菱状长圆形，长圆状披针形或卵状披针形，多少偏斜，长 6～26 cm，宽 3～9 cm，全缘，顶端渐尖或急短渐尖，基部钝圆或短尖、楔形，叶面绿色，无毛，叶背粉绿色或灰白色，有短柔毛，常在中脉和侧脉两侧较密；叶脉在叶背明显，侧脉 8～15 对；小叶柄长 2～10 mm，无毛。花序长 20～36 cm；总花梗、花梗、花萼外面有棕色短绒毛；花淡红色，直径约 4 cm，花梗长 2～3 cm；花萼直径约 2 cm，长 1.2～1.7 cm，顶端具短的 5 齿，内面有疏柔毛或无毛，花瓣阔匙形或倒卵楔形，顶端浑圆，长 1.8～2 cm，宽 1～1.5 cm，无毛，内面有红色纵条纹；花丝长 2.5～3 cm，基部有小柔毛；子房有光亮、白色的柔毛，花柱有柔毛。果椭圆球形，近球形或阔卵形，长 3～5.5 cm，直径 2～3.5 cm，被极短的棕褐色毛和常混生疏白色小柔毛，有或无明显的黄褐色小瘤体，果瓣厚 1.2～5 mm；果柄长 2.5～3.5 cm，有或无毛；种子椭圆球形，平滑，成熟时长约 1.8 cm，直径约 1.3 cm。

【花 果 期】花期 3—9 月，果期 5 月至翌年 4 月。

【分　　布】产于四川、云南、贵州、广西、广东、湖南、湖北、江西、浙江、福建等省（自治区）。越南北部也有分布。

【生　　境】生长于低海拔至中海拔的山地林中。

【保护价值】伯乐树为单种科植物，是古老的残遗种，对研究被子植物的系统发育及古地理等均有科学价值。木材硬度适中，不翘裂，色纹美观，为优良的家具及工艺用材。

【致危因素】生境退化或丧失。

[Scientific Name] *Bretschneidera sinensis*

[Morphological Characters] Trees 10-20 m tall；bark gray-brown；lenticels on branchlets. Pinnate leaves 25-45 cm，sparsely hairy in leaf axils or glabrous；petiole 10-18 cm，pinnate leaflets 7-15，papery or leathery，narrowly elliptic，rhombic-oblong，oblong-lanceolate，or ovate-lanceolate，slightly oblique，6-26×3-9 cm，margin entire，apex acuminate to acute，base obtuse to rounded to cuneate，leaves green adaxially，glabrous，green to glaucous abaxially，with minutely pubescent，densely along midvein and lateral veins；obvious veins abaxially，with lateral veins 8-15 pairs，petiolules 2-10 mm，glabrous. Inflorescence 20-36 cm；pedicel and calyx with brown tomentose；flowers pink，4 cm in diam，pedicel 2-3 cm，calyx 1.2-1.7 cm，2 cm in diam，apex notched 5，inner sparsely pubescent or glabrous，petals broadly spatulate to obovate-cuneate，apex rounded，1.8-2×1-1.5 cm，glabrous，red striate；filaments 2.5-3 cm，hairy at base；ovary and style hairy. Capsule ellipsoid-globose to subglobose to ovoid or obovoid，3-5.5×2-3.5 cm，minutely hairy in brown，with yellowish brown warty，valves 1.2-5 mm thick；fruiting pedicel 2.5-3.5 cm，hairy or glabrous；Seeds ellipsoid-globose，smooth，1.8-1.3 cm when mature.

[Florescence and Fruit Period] Florescence March to September，fruit May to next April.

[Distribution] Sichuan Province，Yunnan Province，Guizhou Province，Guangxi Zhuang Autonomous Region，Guangdong Province，Hunan Province，Hubei Province，Jiangxi Province，Zhejiang Province，Fujian Province and other provinces. It is also distributed in northern Vietnam.

[Habitat] Low to medium altitude mountain forest.

[Protection Value] It is a single family plant and an ancient remnant species，which has scientific value in the study of angiosperm phylogeny and paleogeography. The wood has moderate hardness，no warping，and beautiful color lines. It is a good material for furniture and technology.

[Threatening Factors] Habitat degradation or loss.

摄像：李策宏　陈炳华

Photo：Li Cehong　Chen Binghua

51 膝柄木/ Xibingmu

【学　　名】*Bhesa sinensis*

【形态特征】乔木高达 10 m 以上；小枝粗壮，紫棕色，表面粗糙不平，并常有较大的叶痕和芽鳞痕。叶互生，在小枝上有时近对生，近革质，有光泽，长方窄椭圆形或窄卵形，长 11～20 cm，宽 3.5～6 cm，先端急尖或短渐尖，基部圆形或阔楔形，偶近平截或浅心形，全缘，中脉和侧脉均在叶背强度凸起，侧脉每侧 14～18 对，平行较密排列，三生脉细而多，与中脉略作垂直角度伸出，平行密集成格状网纹；叶柄圆柱状，长 2～3 cm，两端增粗，在近叶片的一端背部微突呈膝状弯曲。花小，黄绿色；聚伞圆锥花序多侧生于小枝上部，常呈假顶生状；花序梗短或近无，花序轴有 3～5 分枝，枝上着生多数短梗小花，如穗状；花 5 数，直径约 5 mm；萼片线状披针形，长约 1.5 mm，先端窄尖；花瓣窄倒卵形或长圆披针形，长约 2 mm，先端圆钝；花盘浅盘状，雄蕊插生其环状外缘上，花丝长约 2 mm，花药内向；子房近扁球状，基部着生花盘上，与之完全游离；子房具 2 心皮，2 室，2 胚珠，上部近花柱处有疏毛丛，花柱 2，粗壮，柱头小。蒴果窄长卵状，长约 3 cm，中下部最宽处宽 1～1.2 cm，上部窄，顶端常稍呈喙状，无小果梗或有长约 1 mm 的粗梗；种子 1 枚，基生，椭圆卵状，长约 1.5 cm，棕红色或棕褐色，有光泽，假种皮淡棕色，包围种子大部，由基部上达种子 2/3 处，先端开口，并有长条状或丝状延伸部分上达种子 4/5 处。

【分　　布】产于广西合浦。印度、越南和马来西亚亦有分布。

【生　　境】生长于海拔 50 m 近海岸的坡地杂木林中。

【保护价值】仅在广西京岛海滩有 15 株左右，面临严重威胁。

【致危因素】自然灾害；物种内在原因。

[Scientific Name] *Bhesa sinensis*

[Morphological Characters] Trees，10 m tall，twigs thick and strong，purple-brown，rough and coarse on surface，often with large leaves and bud scars. Leaves alternate，sometimes subopposite on young twigs，subleathery，glossy，oblong-elliptic or narrowly ovate，11-20×3.5-6 cm，acuminate or tapered at apex，subcuneate or slightly cordate，base often rounded or broadly attenuate，entire margin，main vein and lateral veins prominent abaxially，lateral veins 14-18 pairs，parallel densely，3rd generation veins dense and netted dis-tally，perpendicular to main vein，stretching；petiole cylindric，2-3 cm in length，with thick ends，with abaxially curved part near the leaf blade. Flowers small，yellowish green；cymes on upper part of branchlets，usually pseudoterminal，3-5 branched，many flowered；peduncle short or nearly absent；pedicel short；flowers 5，5 mm in diam；sepals linear-lanceolate，ca. 1.5 mm，narrowly acuminate at apex；petals narrowly obovate or oblong-lanceolate，ca. 2 mm，subrounded at apex；disk in plate shape，with stamens inserted on the outer edge of the ring，filament ca. 2 mm，anther introversion，ovary compressed globose，base surrounded by disk；ovary with 2 carpels，2 chambers，2 ovules，upper part with sparse tufts near style, style 2, stout, with small stigma. Capsule narrow and long ovoid, ca. 3×1-1.2 cm, steeply tapering to apex, sessile, or coarsely pedunculate, ca. 1 mm. Seeds 1, basal, elliptic-ovoid, ca. 1.5 cm, brown or brown-red, glossy; aril covering most of seed, light brown, reaching 2/3 of the seed from the base, apex opened, with elongated or filiform extension reaching 4/5 of the seed.

[Distribution] Hepu, Guangxi Zhuang Autonomous Region. It is also distributed in India, Vietnam and Malaysia.

[Habitat] It grows in the mixed forest on the slope near the coast at an altitude of 50 m.

[Protection Value] There are only about 15 plants in Jingdao beach of Guangxi Zhuang Autonomous Region, which are under serious threat.

[Threatening Factors] Natural disasters; internal causes of species.

摄像：刘 冰 邹 嫦

Photo: Liu Birg Zou Chang

52 萼翅藤/ Echiteng

【学　　名】*Calycopteris floribuunda*

【形态特征】披散蔓生藤本，高 5～10 m 或更高，枝纤细，径约 5 mm，密被柔毛。叶对生，叶片革质，卵形或椭圆形，长 5～12 cm，宽 3～6 cm，先端钝圆或渐尖，基部钝圆，叶面绿色，被柔毛或无毛，主脉及侧脉上被毛，背面密被鳞片及柔毛，侧脉 5～8（10）对，连同网脉在两面明显；叶柄长（8）10（12）mm，密被柔毛。总状花序，腋生和聚生于枝的顶端，形成大型聚伞花序，长 5～15 cm，花序轴被柔毛，苞片浅绿色，脱落；花小，两性；苞片卵形或椭圆形，长 2～3 mm，密被柔毛；花萼杯状，外面被柔毛，长 5～7 mm，5 裂，裂片三角形，长 2～3 mm，直立，两面密被柔毛，外面疏具鳞片；花瓣缺；雄蕊 10 枚，2 轮列，5 枚与花萼对生，5 枚生于萼裂之间，花丝长 2～3 mm，无毛；花药 2 室；子房长 3～4 mm，1 室，胚珠 3 枚，悬垂。假翅果，长约 8 mm，被柔毛，5 棱，萼裂 5 枚，增大，翅状，长 10～14 mm，被毛；种子 1 枚，长 5～6 mm。

【花 果 期】花期 3—4 月，果期 5—6 月。

【分　　布】产于云南盈江。分布于越南、老挝、柬埔寨、马来西亚（槟榔屿）、缅甸、泰国、孟加拉国和印度（德干高原）。

【生　　境】生长于海拔 300～600 m 的季雨林中或林缘地带。

【保护价值】萼翅藤是亚洲热带山地的特有种。它在我国的分布，有力地说明我国云南西部属于热带北缘气候，有一定的科研价值。叶可用作强壮药和去毒药，果用作兴奋剂。

【致危因素】生境退化或丧失。

【附　　注】缺少影像资料。

[Scientific Name] *Calycopteris floribuunda*

[Morphological Characters] Lianas 5-10 m or taller， fine branchlets， 5 mm in diam， densely pilose. Leaves opposite， leaf blades leathery， ovate or ellipse， 5-12×3-6 cm， apex rounded or cuspidate，base rounded，leaflets green adaxially，abaxially densely pubescent or glabrous, persistently pilose on midvein and lateral veins, abaxially densely dark scaly and pilose, lateral veins in 5-8 (10) pairs, obviously on both surfaces with reticulate veins; petiole (8-) 10 (-12) mm, densely pilose. Raceme, axillary and clustered at branchlets apex, forming large panicles, 5-15 cm in length, peduncle and rachis densely pilose, bracts 2-3 mm, ovate or elliptic, densely pilose, light green, glabrescent; flowers small, bisexual; calyx tube 5-7 mm at anthesis, abaxially densely pilose, divided in to 5 triangle lobes, 2-3 mm at anthesis, erect, both surfaces densely pilose, adaxially sparsely scaly; petal with gaps; stamen 10, arranged in 2 rounds, with 5 opposite to calyx, 5 in between calyx lobes, filaments 2-3 mm, glabrous; anthers 2 locular; ovary 3-4 mm long, 1 locular, ovules 3, pendulous. Samara 8 mm, pilose, with 5 ribs, calyx lobes 5, enlarged, winged, 10-14 mm, pilose; seed 1, 5-6 mm long.

[Florescence and Fruit Period] Florescence March to April, fruit May to June.

[Distribution] A native plant of Yingjiang, Yunnan Province. It is distributed in Vietnam, Laos, Cambodia, Malaysia (Penang Island), Myanmar, Thailand, Bangladesh and India (Deccan Plateau).

[Habitat] In or at the edge of monsoon forest at an altitude of 300-600 m.

[Protection Value] It is a unique species of tropical mountains in Asia. Its distribution in China shows that the Western part of Yunnan Province belongs to the climate of the northern margin of the tropics, which has certain scientific research value. Leaves can be used as strong and detoxifying drugs, and fruits as stimulants.

[Threatening Factors] Habitat degradation or loss.

[Note] No image data.

53 革苞菊/ Gebaoju

【学　　名】*Tugarinovia mongolica*

【形态特征】披散蔓生藤本，高 5～10 m 或更高，枝纤细，直径约 5 mm，密被柔毛。叶对生，叶片革质，卵形或椭圆形，长 5～12 cm，宽 3～6 cm，先端钝圆或渐尖，基部钝圆，叶面绿色，被柔毛或无毛，主脉及侧脉上被毛，背面密被鳞片及柔毛，侧脉 5～8（10）对，连同网脉在两面明显；叶柄长 8～10（12）mm，密被柔毛。总状花序，腋生和聚生于枝的顶端，形成大型聚伞花序，长 5～15 cm，花序轴被柔毛，苞片浅绿色，脱落；花小，两性；苞片卵形或椭圆形，长 2～3 mm，密被柔毛；花萼杯状，外面被柔毛，长 5～7 mm，5 裂，裂片三角形，长 2～3 mm，直立，两面密被柔毛，外面疏具鳞片；花瓣缺；雄蕊 10 枚，2 轮列，5 枚与花萼对生，5 枚生于萼裂之间，花丝长 2～3 mm，无毛；花药 2 室；子房长 3～4 mm，1 室，胚珠 3 枚，悬垂。假翅果，长约 8 mm，被柔毛，5 棱，萼裂 5 枚，增大，翅状，长 10～14 mm，被毛；种子 1 枚，长 5～6 mm。

【花 果 期】花期 3—4 月，果期 5—6 月。

【分　　布】产于内蒙古中部。也分布于蒙古南部。

【生　　境】生长于海拔 1 500 m 的干旱草地。

【保护价值】本种为蒙古高原植物区系的特有种，对研究亚洲中部植物区系和菊科植物的系统发育有一定的科学意义。

【致危因素】生境退化或丧失。

【附　　注】缺少影像资料。

[Scientific Name] *Tugarinovia mongolica*

[Morphological Characters] Lianas 5-10 m or taller，fine branchlets，5 mm in diam，densely pilose. Leaves opposite，leaf blades leathery，ovate or ellipse，5-12×3-6 cm，apex rounded or cuspidate，base rounded，leaflets green adaxially，abaxially densely pubescent or glabrous，persistently pilose on midvein and lateral veins，abaxially densely dark scaly and pilose, lateral veins in 5-8 (10) pairs, obviously on both surfaces with reticulate veins; petiole 8-10 (12) mm, densely pilose. Raceme, axillary and clustered at branchlets apex, forming large panicles, 5-15 cm in length, peduncle and rachis densely pilose, bracts 2-3 mm, ovate or elliptic, densely pilose, light green, glabrescent; flowers small, bisexual; calyx tube 5-7 mm at anthesis, abaxially densely pilose, divided in to 5 triangle lobes, 2-3 mm at anthesis, erect, both surfaces densely pilose, adaxially sparsely scaly; petal with gaps; stamen 10, arranged in 2 rounds, with 5 opposite to calyx, 5 in between calyx lobes, filaments 2-3 mm, glabrous; anthers 2-locular; ovary 3-4 mm long, 1-locular, ovules 3, pendulous. Samara 8 mm, pilose, with 5 ribs, calyx lobes 5, enlarged, winged, 10-14 mm, pilose; seed 1, 5-6 mm long.

[Florescence and Fruit Period] Florescence March to April, fruit May to June.

[Distribution] In the middle of Inner Mongolia. Also in Southern Mongolia.

[Habitat] Arid grassland, 1,500 m above sea level.

[Protection Value] This species is endemic to the flora of Mongolia Plateau, which is of scientific significance for the study of flora and phylogeny of Compositae in Central Asia.

[Threatening Factors] Habitat degradation or loss.

[Note] No image data.

54 东京龙脑香/ Dongjing Longnaoxiang

【学　　名】*Dipterocarpus retusus*

【形态特征】大乔木，高约 45 m，具白色芳香树脂；树皮灰白色或棕褐色，不开裂或仅基部纵裂。枝条光滑无毛，具皮孔和较密的环状托叶痕。叶革质，广卵形或卵圆形，长 16～28 cm，宽 10～15 cm，先端短尖，基部圆形或微心形，全缘或中部以上具波状圆齿，上面被白色伏毛，后脱落无毛，下面被疏星状毛，侧脉 16～19 对，下面明显凸起；托叶披针形，长达 15 cm，绿色或红色，无毛（幼树时被星状毛）。总状花序腋生，有 2～5 朵花。花萼裂片 2 枚，较长，为线形，3 枚较短，为三角形；花瓣粉红色，芳香，长椭圆形，长 5～6 cm，先端钝，边缘稍反卷，外面密被鳞片状毛；雄蕊约 30 枚，花药线状披针形，长约 5 mm，基部箭形，相连，药隔附属体芒状，花丝线状，基部扩大为披针形；子房长卵形，被绢状绒毛，3 室，每室具胚珠 2 枚，花柱细柱状，中部以下被长绢毛。坚果卵圆形，密被黄灰色短绒毛；增大的 2 枚花萼裂片为线状披针形，鲜时为红色，长 19～23 cm，宽 3～4 cm，革质，先端圆形，具 3～5 脉，被疏星状短绒毛。

【花 果 期】花期 5—6 月，果期 12 月至翌年 1 月。

【分　　布】产于云南东南部（河口、金平、屏边、绿春、江城）及西部（盈江那帮坝），西藏东南部。印度、缅甸、泰国、老挝、越南、马来西亚、印度尼西亚等有分布。

【生　　境】生长于海拔 1 100 m 以下的潮湿的沟谷雨林及石灰山密林中。

【保护价值】本树种对研究我国热带植物区系和物种资源保存有重要的价值。材质好，出材率高，是优良的用材树种。

【致危因素】生境退化或丧失。

[Scientific Name] *Dipterocarpus retusus*

[Morphological Characters] Trees，lofty，to 45 m tall，with white aromatic resin；bark gray or brown，shallowly flaky，not longitudinally fissured or only at base. Branchlets smooth，glabrous，with lenticels and dense annular stipules. Leaves leathery，broadly ovate or ovate，16-28×10-15 cm，apex with short mucro，base rounded or somewhat cordate，margin entire or sinuate-crenate in distal half，white appressed hair adaxially，glabrescent，adaxially initially white strigose，sometimes glabrescent，abaxially sparsely stellate pubescent，lateral veins 16-19 pairs conspicuously raised abaxially；stipule lanceolate，15 cm，green or red，glabrous.（stellate pubescent when young）. Racemes axillary，2-5 flowered，calyx segments：2 rather long and linear，3 shorter and triangular；petals reddish，sweetly scented，narrowly elliptic，5-6 cm long，apex obtuse，margin slightly reflexed，with dense squamate hairs. Stamens 30；anthers linear to lanceolate，ca. 5 mm，sagittate at base，coherent；connective appendages aristate；filaments filiform，dilated at base. Ovary narrowly ovoid， sericeous-tomentose，3 loculed，with 2 ovules per locule；style finely terete，with long silky hairs on lower half. Nut ovoid，densely yellowish gray tomentellate；winglike calyx segments red，linear-lanceolate，to 19-23×3-4 cm，leathery，stellate-tomentellate，3-5-veined，apex rounded.

[Florescence and Fruit Period] Florescence May to June，fruit period November to January.

[Distribution] Southeast Yunnan Province（Hekou，Jinping，Pingbian，Lyuchun，Jiangcheng）and West Yunnan Province（Yingjiang Nabang dam），Southeast Tibet. Also found in India，Myanmar，Thailand，Laos，Vietnam，Malaysia，Indonesia，etc.

[Habitat] In humid valley rainforests and dense forests of limestone mountain，the altitude is below 1,100 m.

[Protection Value] This species is of great value to the study of tropical flora and conservation of species resources in China. It is a good timber tree with good material and high timber yield.

[Threatening Factors] Habitat degradation or loss.

摄像：朱鑫鑫

Photo：Zhu Xinxin

55 狭叶坡垒/ Xiaye Polei

【学　　名】*Hopea chinensis*

【形态特征】乔木，高 15～20 m，具白色芳香树脂，树皮灰黑色，平滑。枝条红褐色，具白色皮孔，被灰色星状毛或短绒毛。叶互生，全缘，革质，长圆状披针形或披针形，长 7～13 cm，宽 2～4 cm，侧脉 7～12 对，在下面明显凸起，先端渐尖或尾状渐尖，基部圆形或楔形，两侧略不等，上面无毛，下面被疏毛或无毛；叶柄长约 1 cm，黑褐色，具环状裂纹，无毛或被疏毛。圆锥花序腋生、纤细，少花，长 4～18 cm，被疏毛。花萼裂片 5 枚，覆瓦状排列，无毛；花瓣 5 枚，淡红色，扭曲，椭圆形，长 3～4 mm，被黄色长绒毛；雄蕊 15 枚，花药卵圆形，近相等，药隔附属体丝状，长约 2 mm；子房 3 室，每室具胚珠 2 枚。果实卵形，黑褐色，具尖头；增大的 2 枚花萼裂片为长圆状披针形或长圆形，长 8～9 cm，宽 1.5 cm，先端圆形，具纵脉 12 条，无毛。

【花 果 期】花期 6—7 月，果期 10—12 月。

【分　　布】产于广西（十万大山、龙州大青山）。

【生　　境】生长于海拔 600 m 左右的山谷、坡地、丘陵地区。

【保护价值】狭叶坡垒是我国热带季节性雨林的特有珍贵树种。木材材质坚硬，耐腐力极强，有“万年木”之称，可作军工、车船、机械和高级家具等用材。

【致危因素】生境退化或丧失；直接采挖或砍伐。

[Scientific Name] *Hopea chinensis*

[Morphological Characters] Trees 15-20 m tall，with aromatic white resin. Bark grayish black，smooth. Branchlets reddish brown，white lenticellate，gray stellate-pubescent，or gray tomentellate. Leaves alternate，entire margin，leathery，oblong to oblong-lanceolate，7-13×2-4 cm，lateral veins 7-12 pairs，conspicuously raised abaxially，apex acuminate or caudate-acuminate，base rounded or cuneate，slightly asymmetric，adaxially glabrous，sparsely pubescent abaxially；petiole 1 cm，dark brown，with annular cracks，sparsely pubescent or glabrous. Panicle axillary，few flowered，4-18 cm，slender，sparsely pubescent. Sepals 5 imbricate，ovate，glabrous；petals 5 imbricate，reddish，twisted elliptic，3-4 mm，yellow tomentose；stamens 15，anthers ovoid，subequal，connective appendages aristate，ca. 2 mm；ovary 3-loculed，with 2 ovules per locule. Fruit dark brown，ovoid，apiculate；winglike calyx segments slightly swollen，oblong-lanceolate or oblong，8-9× ca. 1.5 cm，apex rounded，longitudinally 12-veined，glabrous.

[Florescence and Fruit Period] Florescence June to July，fruit period October to

December.

[Distribution] Guangxi Zhuang Autonomous Region（Mount Shiwan，Mount Daqing in Longzhou）.

[Habitat] In valley，slope and hilly areas，the altitude is about 600 m.

[Protection Value] It is one of the most valuable tropical seasonal rainforests in China. The wood material is hard and has strong corrosion resistance. It is known as “immortal wood”. It can be used for military industry，vehicles and ships，machinery and advanced furniture.

[Threatening Factors] Degradation or loss of habitat；direct excavation or cutting.

摄像：刘 军 徐晔华 宋 鼎

Photo：Liu Jun Xu Yehua Song Ding

56 坡垒/ Polei

【学　　名】 *Hopea hainanensis*

【形态特征】 乔木，具白色芳香树脂，高约 20 m；树皮灰白色或褐色，具白色皮孔。叶近革质，长圆形至长圆状卵形，长 8～14 cm，宽 5～8 cm，先端微钝或渐尖，基部圆形，侧脉 9～12 对，下面明显突起；叶柄粗壮，长约 2 cm，均无毛或具粉状鳞粃。圆锥花序腋生或顶生，长 3～10 cm，密被短的星状毛或灰色绒毛。花偏生于花序分枝的一侧，每朵花具早落的小苞片 1 枚；花萼裂片 5 枚，覆瓦状排列，长约 2.5 mm，顶端圆形，外面 2 枚全部被毛；花瓣 5 枚，旋转排列，长圆形或长圆状椭圆形，长约 6 mm，宽约 3 mm，先端具不规则的齿缺，基部略收缩偏斜；雄蕊 15 枚，两轮排列，外轮的花丝呈阔卵形，内轮的花丝呈线形，花药卵圆形，药隔附属体丝状，长约 1 mm；子房长圆形，基部具长丝毛，花柱锥状，柱头明显，具花柱基。果实卵圆形，具尖头，被蜡质；增大的 2 枚花萼裂片为长圆形或倒披针形，长 5～7 cm，宽 2.5 cm，具纵脉 9～11 条，被疏星状毛。

【花 果 期】花期 6—7 月，果期 11—12 月。

【分　　布】产于海南。越南北部有分布。

【生　　境】生长于海拔 700 m 左右的密林中。

【保护价值】坡垒是海南热带沟谷雨林的代表种之一，对研究热带植物区系有科研价值。木材坚韧耐久，特别耐水渍，不受虫蛀，为海南树种之冠，适作特种工业、工艺及硬木家具等用。

【致危因素】生境退化或丧失；直接采挖或砍伐。

[Scientific Name] *Hopea hainanensis*

[Morphological Characters] Trees 20 m tall，with aromatic white resin；bark gray or brown，white lenticellate. Leaf blade leathery，oblong to oblong-ovate，8-14×5-8 cm，apex slightly obtuse or acuminate；base rounded，lateral veins 9-12 pairs conspicuously raised abaxially；petiole robust，2 cm long，glabrous or farinose-scurfy. Racemes axillary or apical，3-10 cm long，densely stellate pubescent or gray pubescent. Flowers arranged in lax panicle of unilateral racemes，with 1 bracteole per flower；calyx segments 5，imbricate arranged 2.5 mm，apex rounded outer 2 pubescent on entire surface；petals 5 imbricate，arranged in a spiral，oblong or oblong-elliptic，ca. 6×3 mm，apex irregularly emarginate，base contracted obliquely；stamens 15，arranged in 2 rings，outer ring filaments broad ovate，inner ring filaments liner，anthers ovoid；connective appendages filiform，ca. 1 mm；ovary oblong，hairy at base，style coniform，stigma conspicuous，with stylopodium. Fruit ovoid，apiculate，glaucous；winglike calyx segments oblong or oblanceolate，5-7×ca. 2.5 cm，longitudinally

9-11-veined，sparsely stellate pubescent.

[Florescence and Fruit Period] Florescence June to July，fruit period November to December.

[Distribution] Hainan Province. Also found in Northern Vietnam.

[Habitat] Dense forest about 700 m above sea level.

[Protection Value] Hopea hainanens is one of the representative species of rain forest in the tropical valley of Hainan Province，which is of scientific research value for the study of tropical flora. The wood is tough and durable，especially resistant to water stains and moths. As the top tree species in Hainan Province，it is suitable for special industries，crafts and hardwood furniture.

[Threatening Factors] Degradation or loss of habitat；direct excavation or cutting.

摄像：陈炳华　宋　鼎　张敬莉　徐同波

Photo：Chen Binghua　Song Ding　Zhang Jingli　Xu Tongbo

57 多毛坡垒/ Duomao Polei

【学　　名】*Hopea mollissima*

【形态特征】乔木，高 25～30 m，具白色芳香树脂；树皮灰白色或暗褐色，下部呈不规则的剥落。枝条圆柱形，近叶柄处略有棱，红褐色或灰褐色，被灰色短绒毛或变无毛。叶革质，全缘，长圆形或长圆状披针形，长 8～26 cm，宽 3～8 cm，先端渐尖或尾状渐尖，基部圆形或楔形，两侧略不等，被密毛或星状毛，侧脉 8～11 对，下面明显突起；叶柄长约 1.5 cm，具环状裂纹，密被黄色绒毛。圆锥花序腋生，长 5～9 cm，密被星状毛或疏被毛。花萼裂片 5 枚，覆瓦状排列，卵形，长约 4 mm，宽 2 mm，外面被疏毛，边缘被睫毛；花瓣 5 枚，旋转排列，长 1～1.5 cm，宽 2～4 mm，外面一半（在花蕾时裸露部分）密被绢状绒毛，内面无毛；雄蕊 10～15 枚，两轮排列，花丝基部三角形，花药基着，纵裂，药隔附属体芒状，长约 1.5 mm；子房近卵形，3 室，每室具胚珠 2 枚，花柱短棒状，具明显的花柱基，均无毛。果实卵圆形，径约 1.5 cm，具短尖头，增大的 2 枚花萼裂片线状披针形、长圆形、披针形或倒披针形，长 8～12 cm，宽 1.5～2.5 cm，具疏星状毛，有纵脉 10～13 条。

【花 果 期】花期 7—8 月，果期 12 月至翌年 5 月。

【分　　布】产于云南（屏边、绿春、江城李仙江沿岸）。

【生　　境】生长于海拔 1 000 m 以下潮湿的林中。

【保护价值】毛叶坡垒为我国云南热带雨林的重要组成部分，对阐明云南南部的热带性质有重要意义。木材坚重，是优良的特种建筑用材。

【致危因素】生境退化或丧失；直接采挖或砍伐。

【附　　注】缺少影像资料。

[Scientific Name] *Hopea mollissima*

[Morphological Characters] Trees 25-30 m tall，with aromatic white resin；bark gray or brown，irregular peeling of the lower part. Branches are cylindrical，slightly angulate near the petiole，reddish brown or grayish brown，gray tomentellate，or glabrous. Leaves leathery，entire margin，oblong to oblong-lanceolate，8-26×3-8 cm，apex acuminate or caudate-acuminate，base rounded or cuneate，slightly asymmetric，densely stellate pubescent，lateral veins 8-11 pairs conspicuously raised abaxially；petiole 1.5 cm，with annular cracks，densely yellowish pubescent. Panicle axillary，5-9 cm，sparsely or densely stellate pubescent. Sepals 5，imbricate，ovate，4×2 mm，sparsely tomentose outside，ciliary margin；petals 5，arranged in a spiral，1-1.5 cm×2-4 mm，densely tomentose on parts exposed in bud，glabrous inside；stamen 10-15，arranged in 2 rings，filament base triangular，anther basifixed，with longitudinal cracks，connective appendages aristate，ca. 1.5 mm；ovary 3-loculed，with 2 ovules per locule，style short clavate，with stylopodium，glabrous. Fruit ovoid，1.5 cm in diam.，apiculate，winglike calyx segments linear lanceolate 2，oblong，lanceolate or oblanceolate，8-12×1.5-2.5 cm，stellate-pubescent，longitudinally 10-13 veined.

[Florescence and Fruit Period] Florescence July to August，fruit period December to May.

[Distribution] Yunnan Province（Pingbian，Lyuchun and Lixian River of Jiangcheng）.

[Habitat] In the humid forest below 1,000 m above sea level.

[Protection Value] Hopea mollissima is an important component of the tropical rain forest in Yunnan Province，which is of great significance to clarify the tropical nature of Southern Yunnan Province. The wood is hard and heavy，which is a good special building material.

[Threatening Factors] Habitat degradation or loss；direct excavation or cutting.

[Note] No image data.

58 望天树/ Wangtianshu

【学　名】*Parashorea chinensis*

【形态特征】大乔木，高 40～60 m，胸径 60～150 cm；树皮灰色或棕褐色，树干上部的为浅纵裂，下部呈块状剥落。幼枝被鳞片状的茸毛，具圆形皮孔。叶革质，椭圆形或椭圆状披针形，长 6～20 cm，宽 3～8 cm，先端渐尖，基部圆形，侧脉羽状，14～19 对，在下面明显突起，网脉明显，被鳞片状毛或绒毛；叶柄长 1～3 cm，密被毛；托叶纸质，早落，卵形，基部抱茎，具纵脉 5～7 条，被鳞片状毛或茸毛。圆锥花序腋生或顶生，长 5～12 cm，密被灰黄色的鳞片状毛或绒毛；每个小花序分枝处具小苞片 1 对；每分枝有花 3～8 朵，每朵花的基部具 1 对宿存的苞片，苞片卵形或卵状椭圆形，长 6～13 mm，宽 4～7 mm，具纵脉 6～9 条。花萼裂片 5 枚，覆瓦状排列，长 4～5 mm，宽 1～1.5 mm，内外两面均被鳞片状毛或茸毛；花瓣 5 枚，黄白色，芳香，长 6～11 mm，宽 3～7 mm，具纵脉 10～14 条，外面被鳞片状毛，内面近无毛；雄蕊 12～15 枚，长约 4 mm，两轮排列，花药线状披针形，药室顶部具尖头，药隔附属体锥状；子房长卵形，3 室，每室具胚珠 2 枚，密被白色的绢状毛，花柱细柱状，无毛，柱头小，略 3 裂。果实长卵形，密被银灰色的绢状毛；果翅近等长或 3 长 2 短，近革质，长 6～8 cm，宽 0.6～1 cm，具纵脉 5～7 条，基部狭窄不包围果实。

【花 果 期】花期 5—6 月，果期 8—9 月。

【分　　布】产于云南勐腊、河口，广西那坡、巴马、龙州等。

【生　　境】生长于海拔 300～1 100 m 的沟谷、坡地、丘陵及石灰山密林中。

【保护价值】望天树树干高大而通直，材质优良，加工性能良好，是热带优良的用材树种。同时对研究我国的热带植物区系有重要意义。

【致危因素】直接采挖或砍伐。

[Scientific Name] *Parashorea chinensis*

[Morphological Characters] Trees evergreen，lofty emergents，40-60 m tall，0.6-1.5 m in diam；bark gray or brown，shallowly longitudinally fissured on upper part，but exfoliating in masses on lower part. Branchlets scurfy pubescent or tomentose，lenticels rounded. Leaf blade leathery，elliptic-lanceolate，6-20×3-8 cm，apex acuminate，base rounded，lateral veins pinnate in 14-19 pairs，conspicuously raised abaxially，tertiary veins conspicuous and subreticulate，scurfy pubescent or tomentose；petiole 1-3 cm long，densely pubescent；stipules papery，caducous，ovate，basally amplexicaul，with 5-7 longitudinal veins，scurfy pubescent or tomentose；panicles axillary or terminal，5-12 cm long，densely covered with grayish yellow scaly hairs or villi；cymes 2-bracteate at base，3-8-flowered. Pedicel 2-bracteolate；bracts and bracteoles ovate or ovate-elliptic，6-13×4-7 mm，longitudinally 6-9-veined. Calyx segments 5，imbricate，4-5×1-1.5 mm，scurfy pubescent or tomentose on both surfaces；petals 5，yellowish white，sweet fragrance，6-11×3-7 mm，longitudinally 10-14-veined，scurfy pubescent outside，glabrous inside；Stamens 12-15，4 mm long，arranged in 2 rings，anthers linear-lanceolate，outer cells apiculate，connective appendages acicular；ovary narrowly

ovoid, 3-loculed, with 2 ovules per locule, white sericeous-tomentose, style finely terete, glabrous, stigma small, slightly 3-lobed. Fruit ellipsoid, densely silvery silky-pubescent; calyx segments subequal, or 3 long and 2 short, winglike, 6-8 cm×6-10 mm, longitudinally 5-7-veined, bases narrow, not completely enveloping fruit.

[Florescence and Fruit Period] Florescence May to June, fruit period August to September.

[Distribution] Yunnan Province (Mengla, Hekou), Guangxi Zhuang Autonomous Region (Napo, Bama, Longzhou, etc.).

[Habitat] It grows in the dense forests of valleys, slopes, hills and limestone mountains, with an altitude of 300-1,100 m.

[Protection Value] The trunk of *Parashorea chinensis* is tall and straight, the material is suitable for processing, and it is a good timber tree species in the tropics. At the same time, it is of great significance to the study of tropical flora in China.

[Threatening Factors] Direct digging or cutting.

摄像：宋　鼎　徐晔春

Photo: Song Ding　Xu Yechun

59 貉藻/ Hezao

【学　　名】*Aldrovanda vesiculosa*

【形态特征】浮水草本，长 6～10 cm。叶轮生，每轮 6～9 片，基部合生；叶柄长 3～4 mm，顶部具 4～6 条钻形裂条，裂条长 5～7 mm，叶片平展时肾状圆形，长 4～6 mm，宽 6～10 mm，具腺毛和感应毛，受刺激时两半以中肋为轴互相靠合，外圈紧贴，中央形成一囊体，以此捕捉昆虫。花单生叶腋，具短柄；萼片 5 枚，基部合生，卵状椭圆形或椭圆状长圆形，长 3～4 mm；花瓣 5 片，白色或淡绿色，长圆形，长 3～4 mm，宽约 2.5 mm；雄蕊 5 枚，花丝钻形，花药纵裂；子房近球形，直径 2～2.5 mm，侧膜胎座 5 枚，花柱 5 枚，顶部扩大多裂。果实近球形，不开裂；种子由果皮腐烂而出，5～8 粒或更少，黑色。

【花 果 期】花期未知，果期 8 月。

【分　　布】产于黑龙江。分布于欧洲中部、南部，亚洲北部和东南部，大洋洲北部。

【生　　境】生长于水中。

【保护价值】科研。

【致危因素】生境退化或丧失。

【附　　注】缺少影像资料。

[Scientific Name] *Aldrovanda vesiculosa*

[Morphological Characters] Floating herb，stem 6-10 cm，leaf whorls 6-9-leaved；basal growth；petiole 3-4 mm，apex with 4-6 bristles，bristles 5-7 mm，extending slightly beyond leaf blade，reniform-orbicular，4-6×6-10 mm，with glandular hairs and sensory hair，when stimulated，the two halves lean against each other with the central rib as the axis，outer ring close to each other，forming a capsule in the center to catch insects. Flowers solitary in axils of leaves，shortly stipitate；sepals 5，connate at base，ovate elliptic or elliptic oblong，3-4 mm；petals 5，white or light green，oblong，3-4×2.5 mm；stamens 5，filaments subulate，anthers longitudinally dehiscent; ovary subglobose, 2-2.5 mm in diam., lateral membranous placentation 5, style 5, apical enlargement multifractous. Fruit subglobose, uncracked; seeds rotted from pericarp, 5-8 or less, black.

[Florescence and Fruit Period] Florescence is unknown, fruit is August.

[Distribution] Heilongjiang Province. It is distributed in Central and Southern Europe, North and Southeast Asia and North Oceania.

[Habitat] In water.

[Protection Value] Scientific research.

[Threatening Factors] Habitat degradation or loss.

[Note] No image data.

60 瑶山苣苔/ Yaoshan Jutai

【学　　名】 *Dayaoshania cotinifofia*

【形态特征】 多年生草本。根状茎近圆柱形，顶端粗约 7 mm。叶 9～17 枚，均基生；叶片纸质，宽椭圆形、圆卵形或近圆形，长 2.5～5.5 cm，宽 2.3～4.8 cm，顶端微尖、钝或圆形，基部稍斜，圆形或宽楔形，偶尔浅心形，边缘近全缘或有不明显小浅钝齿，两面稍密被白色短柔毛，侧脉每侧 4～7 条，下面稍隆起；叶柄长 0.8～6 cm，密被贴伏短柔毛。聚伞花序 2～4 条，每花序有 1～2 花；花序梗长 5.5～8.5 cm，与花梗均密被短柔毛；苞片对生，线状披针形，长 5.5～9 mm，宽 1.2～2.2 mm，密被短柔毛；花梗长 4～12 mm。花萼 5 枚全裂，裂片狭三角形或披针状线形，长 5～8 mm，宽 1.2～2 mm，边缘近全缘或有少数小齿，有 3 条脉，外面稍密被短柔毛，内面只在上部疏被短柔毛。花冠淡紫色或白色，长 1.3～1.9 cm，外面疏被短柔毛；筒长 7～9 mm，口部直径 5～9 mm，内面疏被短柔毛；檐部直径 1～2 cm，上唇长 7～10 mm，2 裂，裂片宽卵形或圆卵形，宽 6～10 mm，下唇长 7～12 mm，2～3 裂近中部，裂片三角形，宽 2～6 mm，边缘有短柔毛。雄蕊 1～2；花丝着生于距花冠基部 1.8～2 mm 处，狭线形，长 7.5～14 mm，疏被短柔毛，花药暗紫色，长圆形，长 2～3 mm，无毛；退化雄蕊 2 或不存在，狭线形，长约 4 mm，被短柔毛。花盘环状，高约 1 mm。雌蕊长 1～1.6 cm，子房线形，长 4.5～9 mm，密被短柔毛，花柱疏被短柔毛，柱头半圆形或宽卵形，宽约 0.6 mm。幼果线形，长约 2.5 cm，被短柔毛。

【花 果 期】 花期 9 月。

【分　　布】 产于广西金秀大瑶山。

【生　　境】生长于海拔 860～1 200 m 的山地林中或路边林下。
【保护价值】数量稀少；珍贵；特有。
【致危因素】直接采挖或砍伐；物种内在因素。

[Scientific Name] *Dayaoshania cotinifofia*

[Morphological Characters] Herbs perennial. Rhizome nearly cylindrical, apex ca. 7 mm thick. Leaves 9-17, all basal; leaves papery, broadly elliptic, ovate or nearly rounded, 2.5-5.5×2.3-4.8 cm wide, apex slightly acuminate, blunt or rounded, base slightly oblique, broadly cuneate to nearly cordate, margin nearly entire to indistinctly crenulate, densely white puberulent on both surfaces, Lateral veins 4-7 on each side, slightly raised abaxially; petiole 0.8-6 cm long, densely appressed pubescence. Cymes 2-4, flowers 1-2 per inflorescence; peduncle 5.5-8.5 cm, densely pubescent with peduncle; bracts opposite, linear lanceolate, 5.5-9×1.2-2.2 mm, densely pubescent; peduncle 4-12 mm. Calyx 5-lobed, lobes narrowly triangular or lanceolate linear, 5-8×1.2-2 mm, margin entire or with a few denticles, with 3 veins, outside slightly densely pubescent, inside only sparsely pubescent in the upper part. Corolla purplish or white, 1.3-1.9 cm, sparsely pubescent on the outside; tube 7-9 mm, mouth 5-9 mm in diam, sparsely pubescent on the inside; eaves 1-2 cm in diam, upper lip 7-10 mm, 2-lobed, lobes broadly ovate or oval, 6-10 mm, lower lip 7-12 mm, 2-3-lobes near the middle, lobes triangular, 2-6 mm wide, margin pubescent. Stamens 1-2; filaments inserted 1.8-2 mm from the base of corolla, narrowly linear, 7.5-14 mm long, sparsely pubescent, anthers dark purple, oblong, 2-3 mm, glabrous; staminodes 2 or absent, narrowly linear, about 4 mm, pubescent. Disk annular, ca. 1 mm tall. Pistil 1-1.6 cm, ovary linear, 4.5-9 mm long, densely pubescent, style sparsely pubescent, stigma semicircular or broadly ovate, about 0.6 mm. Young fruit linear, ca. 2.5 cm, pubescent.

[Florescence and Fruit Period] Florescence September.

[Distribution] Mount Dayao, Jinxiu, Guangxi Zhuang Autonomous Region.

[Habitat] In mountain forest or roadside forest, 860-1,200 m above sea level.

[Protection Value] Rare in quantity; precious; peculiar.

[Threatening Factors] Direct digging or felling; internal factors of species.

摄像：朱鑫鑫

Photo: Zhu Xinxin

61 单座苣苔/ Danzuo Jutai

【学　　名】*Metabriggsia ovalifolia*

【形态特征】多年生草本。茎高 20～40 cm，被褐色长柔毛。叶具短或长柄；叶片草质，两侧稍不对称，卵形，长 5～25.5 cm，宽 2.5～17 cm，顶端渐尖，基部斜圆形，边缘有浅波状小钝齿，两面被贴伏短柔毛，下面脉上及边缘有密的柔毛，侧脉每侧 5～10 条，与中脉成钝角或直角展出，在近边缘时向上弯曲；叶柄长 0.3～7 cm，被与茎相同的毛。聚伞花序生于茎上部叶腋，有长梗，有 5～12 花，2～3 回分枝，节膨大；花序梗长 7.5～12.5 cm，被褐色腺毛；花梗长 5～6 mm，被短柔毛。花萼裂片长 9～10 mm，宽 1.5～2 mm，顶端微钝，外面被短柔毛，内面无毛，有 3～5 条脉。花冠白色，带黄绿色，长约 3.6 cm，外面被疏柔毛，内面无毛；筒长约 2.7 cm，口部直径 1.5 cm，基部直径 5 mm，下面稍膨胀；上唇长 2.8 mm，宽 2.9 cm，2 裂，裂片斜正三角形，顶端圆形，下唇长 10 mm，宽 12 mm，3 浅裂，中裂片宽卵形，侧裂片斜三角形。雄蕊无毛，花丝着生于距花冠基部 1.4 cm 处，长 1.4 cm，花药长 3.5 mm；退化雄蕊 3 枚，无毛，侧生 2 枚着生于距花冠基部 6 mm 处，长 9～10 mm，顶端稍变宽，中央 1 枚小，着生于距花冠基部 7 mm 处，长 1.5 mm。花盘高约 1 mm，边缘波状。雌蕊长约 2.5 cm，子房长约 8 mm，宽 1 mm，疏被短柔毛，花柱长 16 mm，下部疏被短柔毛，柱头小。蒴果线形，长约 1.5 cm，粗 3 mm，被短柔毛。种子长约 0.4 mm。

【花 果 期】花期 10 月。

【分　　布】产于广西那坡。

【生　　境】生长于海拔 1 100 m 的石灰山山坡林下。
【保护价值】数量稀少；珍贵；特有。
【致危因素】直接采挖或砍伐；物种内在因素。

[Scientific Name] *Metabriggsia ellipsoidifolia*

[Morphological Characters] Perennial herb, stems 20-40 cm, brown villous. Leaves with short or long stalks, papery, slightly asymmetric on both sides, ovate, 5-25.5×2.5-17 cm, apex acuminate, base obliquely round, margin repand-crenulate, shortly pubescent on the surfaces, densely pubescence on veins and margin abaxially. Lateral veins 5-10 on each side, obtuse or at right angles to midvein, curved upward near margin; petiole 0.3-7 cm long, covered with the same hairs as stem. Cymes in axils of upper leaves, with long peduncles, flowers 5-12, branches 2-3, swollen nodes; peduncles 7.5-12.5 cm, covered with brown glandular hairs; peduncles 5-6 mm, covered with pubescence. Calyx lobes 9-10×1.5-2 mm, apex obtuse, outside pubescent, inside glabrous, with 3-5 veins. Corolla white, suffused yellow-green, ca. 3.6 cm, outside sparsely pubescent near apex, inside glabrous; tube ca. 2.7 cm; mouth 1.5 cm in diam, base 5 mm in diam, lower part slightly expanded; adaxial lip ca. 2.8 mm×2.9 cm, divided into 2 triangle lobes, apex rounded, abaxial lip ca. 10×12 mm, divided into 3 triangle lobes, middle lobe broadly ovate, lateral lobes triangle. Stamens glabrous, filaments inserted at 1.4 cm from corolla base, 1.4 cm long, anthers 3.5 mm; staminodes 3, glabrous, lateral 2 at 6 mm from corolla base, 9-10 mm, apex slightly wider, central 1 small, 7 mm from the base of corolla, 1.5 mm in length. Disk ca. 1 mm tall, margin undulate. Pistil ca. 2.5 cm; ovary ca. 8×1 mm, sparsely puberulent, style 16 mm, lower part sparsely pubescent, stigma small. Capsule linear, ca. 1.5 cm, ca. 3 mm thick, pubescent. Seeds 0.4 mm.

[Florescence and Fruit Period] Florescence October.

[Distribution] Napo, Guangxi Zhuang Autonomous Region.

[Habitat] In the forest on the hillside of limestone mountain, 1,100 m above sea level.

[Protection Value] Rare in quantity; precious; peculiar.

[Threatening Factors] Direct digging or felling; internal factors of species.

摄像：王明川

Photo: Wang Mingchuan

62 报春苣苔/ Baochun Jutai

【学　　名】*Primulina tabacum*

【形态特征】多年生草本，有烟草气味。叶均基生，具长或短柄；叶片圆卵形或正三角形，长 5～10 cm，顶端微尖，基部浅心形，边缘浅波状或羽状浅裂，裂片扁正三角形，两面均被短柔毛，下面还有腺毛，侧脉每侧约 3 条，上面平，下面稍隆起；叶柄长 2.5～14 cm，扁平，边缘有波状翅。聚伞花序伞状，1～2 回分枝，有 3～9 花；花序梗与叶等长或比叶短，被短柔毛和短腺毛；苞片对生，狭长圆形或线状披针形，长 1.5 cm，有腺毛。花萼长约 6.5 mm，5 深裂，两面被短柔毛，筒长约 1 mm；裂片狭披针形或条状披针形，长约 5.5 mm，宽 0.8～1.1 mm，顶端有腺体，边缘上部每侧有 1～2 个三角形小齿，齿顶端有腺体。花冠紫色，外面和内面均被短柔毛；细筒状，长约 9 mm，口部直径 3 mm；檐部平展，直径约 1.6 cm，不明显二唇形，上唇长约 7 mm，2 深裂，裂片狭倒卵形，长约 5 mm，宽 3.2 mm，顶端钝，下唇长约 9 mm，3 深裂，裂片也为狭倒卵形，长约 6 mm，宽 4 mm，顶端圆形。雄蕊无毛，花丝着生于距花冠基部约 1 mm 处，近丝形，长约 0.8 mm，花药长圆形，长约 1.5 mm，连着；退化雄蕊 3，长 0.2～0.3 mm。花盘高约 0.5 mm，由 2 近方形腺体组成。雌蕊长约 2.6 mm，子房狭卵形，长约 1.5 mm，与花柱被短柔毛，花柱粗，长约 0.5 mm，柱头长约 0.6 mm，2 浅裂。蒴果长椭圆球形，长 3.2～6 mm。种子暗紫色，狭椭圆球形，长约 0.4 mm，有密集小乳头状突起。

【花 果 期】花期 8—10 月。

【分　　布】产于广东北部连县、阳山一带。

【生　　境】溶洞植物。

【保护价值】数量稀少；珍贵；特有；受威胁严重；科学及文化价值大。

【致危因素】生境退化或丧失。

[Scientific Name] *Primulina tabacum*

[Morphological Characters] Perennial herb, with smell of tobacco. Leaves basal, long or short stipe; leaf blade oval or triangle, 5-10 cm, apex slightly acuminate, base lightly cardioid, margin lightly undulate or pinnate, lobes flat triangle, pubescent on both surfaces, abaxial glandular hairs, lateral veins 3 on each surface, flat adaxially, slightly raised abaxially; petiole 2.5-14 cm, flat, margin undulate winged. Cymes 1-2-branched, 3-9-flowered; peduncles as long as leaves or shorter than leaves, pubescent and short glandular hairs; bracts opposite, narrowly oblong or linear lanceolate, 1.5 cm long, with glandular hairs. Calyx 6.5 mm long, 5-deeply split, pubescent on both surfaces, tube 1 mm in length; lobes narrow lanceolate or liner lanceolate, 5.5×0.8-1.1 mm, with glands at the top, 1-2 triangular denticles on each side of the upper edge, with glands at the top of the teeth. Corolla purple, pubescent on the outside and inside; thin tube, 9 mm in length, mouth 3 mm in diam; leaves flat, 1.6 cm in diam, not obvious double-lip shaped, adaxial lip about 7 mm long, 2 deep split, lobes narrow obovate, 5×3.2 mm, blunt top, abaxial lip 9 mm, 3 deep split, lobes narrow obovate, 6×4 mm, apex rounded. Stamens glabrous, filaments inserted 1 mm from the base of corolla, nearly filiform, 0.8 mm long, anthers oblong, 1.5 mm long, contiguous; staminodes 3, 0.2-0.3 mm. Disk ca. 0.5 mm, composed of 2 nearly square glands. Pistil 2.6 mm, ovary narrowly ovate, 1.5 mm long, pubescent with style, style thick, 0.5 mm, stigma about 0.6 mm, 2-lobed. Capsule ellipsoid, 3.2-6 mm. Seeds dark purple, narrowly ellipsoid, ca. 0.4 mm, with dense papillae,

[Florescence and Fruit Period] Florescence August to October.

[Distribution] Lianxian and Yangshan in the north of Guangdong Province.

[Habitat] Cave plants.

[Protection Value] Rare, precious, unique, seriously threatened, of great scientific and cultural significance.

[Threatening Factors] Habitat degradation or loss.

摄像：周建军　周　辉　刘　昂

Photo: Zhou Jianjun　Zhou Hui　Liu Ang

63 辐花苣苔/ Fuhua Jutai

【学　　名】*Thamnocharis esquirolii*

【形态特征】多年生小草本。根状茎短。叶 14～18 片，均基生，具柄；叶片纸质，多椭圆形，稀狭倒卵形，长 1.2～5 cm，宽 0.7～2.8 cm，顶端微尖或钝，基部楔形或宽楔形，边缘有小钝齿，两面密被贴伏的白色短柔毛，侧脉每侧 3～4 条，叶柄长 0.6～4 cm。聚伞花序约 3 条，每花序有 5～9 花；花序梗长 3～5 cm，与花梗均密被短柔毛；苞片对生，极小，钻形，长 1.5～2 mm，被短柔毛；花梗长 0.6～4 mm。花萼钟状，长 2.2～3 mm，4～5 裂近基部，裂片稍不等大，三角形，宽 0.7～1.1 mm，外面被短柔毛，内面无毛。花冠紫色或蓝色，辐状，4～5 深裂，直径约 12 mm；筒长约 2 mm，无毛；裂片披针状长圆形，长 6～7 mm，宽 2～3 mm，顶端微钝，外面上部被短柔毛，内面无毛。雄蕊 4～5 枚，不等长，花丝长 2.5～7 mm，宽 0.2～0.3 mm，疏被短柔毛，花药宽椭圆形，长 1.3～1.8 mm，无毛。花盘小，高 0.2 mm。雌蕊长约 5 mm，子房卵球形，长 2 mm，被短柔毛，花柱长 3 mm，无毛，柱头近截形。蒴果线状披针形，长约 11 mm，宽 2.8 mm，疏被糙伏毛。

【花 果 期】花期 8 月。

【分　　布】特产于贵州兴仁及贞丰一带。

【生　　境】生长于海拔 1 500～1 600 m 的山地灌丛中或林下。

【保护价值】数量稀少，珍贵，特有，受威胁严重，科学及文化价值大。

【致危因素】生境退化或丧失。

[Scientific Name] *Thamnocharis esquirolii*

[Morphological Characters] Small perennial herbs. Rhizome short. Leaves 14-18, basal, stipitate; leaf blade papery, usually elliptic, thin and obovate, 1.2-5×0.7-2.8 cm, apex tip is slightly acuminate or blunt, base wedge-shaped or wide wedge-shaped, margin shallowly crenulate to serrulate, densely white pubescent on both surfaces, lateral veins 3-4 on each surface, with petiole 0.6-4 cm. Cymes 3 with 5-9 flowers in each inflorescence; inflorescence peduncle 3-5 cm, densely pubescent with the peduncle; bracts opposite, tiny, subulate, 1.5-2 mm, pubescent; peduncle 0.6-4 mm. Calyx campanulate, 2.2-3 mm, with 4-5-lobed near base, lobes slightly unequal, triangular, 0.7-1.1 mm in width, pubescent outside, glabrous inside. Corolla purple or blue, radial, 4-5-parted, ca. 12 mm in diameter; tube ca. 2 mm, glabrous; lobes lanceolate oblong, 6-7×2-3 mm, apex obtuse, outer upper part pubescent, inner surface glabrous. Stamens 4-5, unequal, filaments 2.5-7×0.2-0.3 mm, sparsely pubescent, anthers broadly elliptic, 1.3-1.8 mm long, glabrous. Disk small, 0.2 mm. Pistil 5 mm, ovary ovoid, 2 mm, pubescent, style 3 mm, glabrous, stigma nearly truncate. Capsule linear lanceolate, ca. 11×2.8 mm, sparsely strigose.

[Florescence and Fruit Period] Florescence August.

[Distribution] A native plant of Xingren and Zhenfeng, Guizhou Province.

[Habitat] In or under mountain shrubs, 1,500-1,600 m above sea level.

[Protection Value] Rare; precious; unique; seriously threatened; of great scientific and cultural significance.

[Threatening Factors] Habitat degradation or loss.

摄像：刘　冰

Photo: Liu Bing

64 华山新麦草/ Huashan Xinmaicao

【学　　名】*Psathyrostachys huashanica*

【形态特征】多年生，具延长根茎。秆散生，高 40～60 cm，直径 2～3 mm。叶鞘无毛，基部褐紫色或古铜色，长于节间；叶舌长约 0.5 mm，顶具细小纤毛；叶片扁平或边缘稍内卷，宽 2～4 mm，分蘖者长 10～20 cm，秆生者长 3～8 cm，边缘粗糙，上面黄绿色，具柔毛，下面灰绿色，无毛。穗状花序长 4～8 cm，宽约 1 cm；穗轴很脆，成熟时逐节断落，节间长 3.5～4.5 mm，侧棱具硬纤毛，背腹面具微毛；小穗 2～3 枚生于 1 节，黄绿色，含 1～2 小花；小穗轴节间长约 3.5 mm；颖锥形，粗糙，长 10～12 mm；外稃无毛，粗糙，第一外稃长 8～10 mm，先端具长 5～7 mm 的芒；内稃等长于外稃，具 2 脊，脊上部疏生微小纤毛；花药黄色，长约 6 mm。

【花 果 期】花果期 5—7 月。

【分　　布】特产于我国陕西华山。

【生　　境】生长于山坡道旁岩石残积土上。

【保护价值】科研。

【致危因素】生境退化或丧失；直接采挖或砍伐。

[Scientific Name] *Psathyrostachys huashanica*

[Morphological Characters] Plants perennial, laxly tufted, with long, underground rhizomes. Culms glabrous, 40–60 cm tall, 2-3 mm in diam. Leaf sheath glabrous, base brown purple or bronze, growing between nodes; ligule 0.5 mm, apex with fine cilia; leaf blade flat or slightly involute at the margin, 2-4 mm wide, tiller 10-20 cm long, culm 3-8 cm, margin scabrous, yellow green on the top, pilose, gray green below, glabrous. Spike 4-8×1 cm; rachis fragile, falling off when mature, internode 3.5-4.5 mm, hard cilia on the lateral edge, sparsely pubescent abaxially; spikelets 2 or 3 per node, yellowish green, with 1 or 2 florets; rachilla internodes 3.5 mm; glumes 10-12 mm, tapered, scabrous; lemma glabrous, scabrous, first lemma 8-10 mm, with awn 5-7 mm; palea equaling lemma, with 2 keels, ciliolate along keels, anthers yellow, 6 mm.

[Florescence and Fruit Period] Florescence and fruit period May to July.

[Distribution] A native plant of Mount Hua, Shaanxi Province.

[Habitat] On the rock residual soil beside the hillside road.

[Protection Value] Scientific research.

[Threatening Factors] Degradation or loss of habitat; direct digging or cutting.

摄像：李智选

Photo: Li Zhixuan

65 银缕梅/ Yinlyumei

【学　　名】 *Shaniodendron subaequalum*

【形态特征】 落叶小乔木。芽及幼枝被星状毛。叶互生，薄革质，椭圆形或倒卵形，长 5～9 cm，先端尖，基部不等侧圆，两面被星状毛，具不整齐粗齿；叶柄长 4～6 mm，被星状毛，托叶披针形，早落。短穗状花序腋生及顶生，具 3～7 花；雄花与两性花同序，外轮 1～2 朵为雄花，内轮 4～5 朵为两性花。花无梗，苞片卵形；萼筒浅杯状，萼具不整齐钝齿，宿存；无花瓣；雄蕊 5～15 枚，花丝长，直伸，花后弯垂，花药 2 室，具 4 个花粉囊，药隔突出；子房半下位，2 室，花柱 2 枚，常卷曲。蒴果木质，长圆形，长 1.2 cm，被毛，萼筒宿存果及萼筒均密被黄色星状柔毛。种子光褐色。

【保护价值】 数量稀少；珍贵；特有；受威胁严重；科学及文化价值大。

【致危因素】 直接采挖或砍伐；生境退化或丧失；种间影响。

[Scientific Name] *Shaniodendron subaequalum*

[Morphological Characters] Small deciduous trees. Buds and young branchlets with stellate hairy. Leaves alternate，thinly leathery，elliptic or obovate，5-9 cm，apex acute，base unequally lateral rounded，both surfaces stellate hairy，with irregular coarse teeth；petiole 4-6 mm，stellate hairy，stipules lanceolate，caducous. Short spikes axillary and terminal，with 3-7 flowers；male flowers and bisexual flowers in the same inflorescence，with 1-2 male flowers in the outer ring and 4-5 male flowers in the inner ring. Flowers sessile，bracts ovate；calyx tube shallowly cup-shaped，calyx with irregular blunt teeth，persistent；without petals；stamens 5-15，filaments long，straight，curved after anther，anthers 2-celled，with 4 pollen sacs，anther septum prominent；ovary half inferior，2-celled，style 2，often curled. Capsule woody，oblong，1.2 cm long，hairy，calyx tube persistent fruit and calyx tube densely yellow stellate pilose. Seeds light brown.

[Protection Value] Rare，precious，unique，seriously threatened，of great scientific and cultural significance.

[Threatening Factors] Direct digging or cutting；habitat degradation or loss；interspecific impact.

摄像：潘　哲　王启平

Photo：Pan Zhe　Wang Qiping

66 长蕊木兰/ Changrui Mulan

【学　　名】*Alcimandra cathcardii*

【形态特征】乔木，高达 50 m，胸径达 50 m；嫩枝被柔毛；顶芽长锥形，被白色长毛。叶革质，卵形或椭圆状卵形，长 8～14 cm，先端渐尖或尾状渐尖，基部圆或阔楔形，上面有光泽，侧脉每边 12～15 条，纤细，末端与密致的网脉网结而不明显，叶柄长 1.5～2 cm，无托叶痕。花白色，佛焰苞状苞片绿色，紧接花被片，花梗长约 1.5 cm；花被片 9，有透明油点，具约 9 条脉纹，外轮 3 片长圆形，长 5.5～6 cm，宽 2～2.2 cm；内两轮倒卵状椭圆形，比外轮稍短小，药隔伸长成短尖；雄蕊长约 4 cm，花药长约 2.8 cm，内向开裂；雌蕊群圆柱形，长约 2 cm，直径 3 mm，具约 30 枚雌蕊，雌蕊群柄长约 1 cm。聚合果长 3.5～4 cm；蓇葖扁球形，有白色皮孔。

【花 果 期】花期 5 月，果期 8—9 月。

【分　　布】产于云南西南部至东南部、西藏南部和东南部。常与壳斗科、樟科树种混交成林。印度（阿萨姆）也有分布。

【生　　境】生长于海拔 1 800～2 700 m 的山林中。

【致危因素】直接采挖或砍伐；生境退化或丧失。

[Scientific Name] *Alcimandra cathcardii*

[Morphological Characters] Trees, up to 50 m in height and 50 m d.b.h; shoots pilose;

terminal buds long conical, white long villous. Leaves leathery, ovate or oval ovate, 8-14 cm long, apex acuminate or caudate acuminate, base round or broad wedge-shaped, glossy on the top, lateral veins 12-15 on each surface, slender, base not obvious with dense network of network veins, petiole 1.5-2 cm, without stipule mark. Flowers white, spathe like bracts green, followed by perianth pieces, pedicel 1.5 cm; perianth pieces 9, with transparent oil spots, with about 9 veins, outer round 3 pieces, 5.5-6×2-2.2 cm; inner two rounds obovate elliptic, slightly shorter than outer round, connective extension into short tip; stamens 4 cm, anthers 2.8 cm, introrse cracking; pistil group cylindrical, 2 cm×3 mm, with 30 pistils, pistil group handle about 1 cm long. Fruit 3.5-4 cm; follicles ellipsoid, with white lenticels.

[Florescence and Fruit Period] Florescence May, fruit period August to September.

[Distribution] Southwest to Southeast of Yunnan Province, South and Southeast of Tibet. It is often mixed with Fagaceae and camphoridae species. Also found in India（Assam）.

[Habitat] In the mountain forest with an altitude of 1,800-2,700 m.

[Threatening Factors] Direct digging or cutting; habitat degradation or loss.

摄像：李策宏

Photo：Li Cehong

67 单性木兰/ Danxing Mulan

【学　　名】*Alcimandra cathcardii*

【形态特征】乔木，高达 18 m，胸径 40 cm，树皮灰色；小枝绿色，初被平伏短柔毛。叶革质，椭圆状长圆形或倒卵状长圆形，长 8～15 cm，宽 3.5～6 cm，先端圆钝而微缺，基部阔楔形，叶两面无毛，或叶背嫩时基部有稀疏柔毛，上面亮绿色，侧脉每边 12～17 条，干时网脉两面凸起，叶柄长 2～3.5 cm，初被灰色柔毛，后变无毛，托叶痕几达叶柄先端。花单性异株，雄花：花被片白带淡绿色，外轮 3 片倒卵形，长 2～3 cm，宽约 2 cm，内轮 2 片，椭圆形，稍狭小；雄蕊群白色带淡黄色，倒卵圆形；雄蕊长 1.8～2.5 cm，花药长约 1.5 cm，侧向开裂，药隔伸出长 2～3 mm 的舌状尖头，花丝长约 2 mm；雌花：外轮花被片 3，倒卵形，长 2.5～3 cm，宽 2～2.5 cm，内轮花被片 8～10，线状披针形，长 2～2.5 cm，宽 0.2～0.3 cm；雌蕊群绿色，倒卵圆形，长约 1 cm，径 0.8～0.9 cm，具 6～9 枚雌蕊，花柱短，柱头面鸡冠状，每心皮具胚珠 2 枚，胚珠有短柄。聚合果近球形，果皮革质，熟时红色，径 3.5～4 cm，蓇葖背缝全裂，具种子 12 枚；种子外种皮红色，豆形或心形，宽 10～12 mm，高 7～9 mm，去外种皮种子黑褐色，顶端平或稍凹，具狭长沟，中具柄，腹背两面数块具不规则凸起。

【花 果 期】花期 5—6 月，果期 10—11 月。

【分　　布】产于广西北部（罗城、环江）、贵州东南部（荔波）。

【生　　境】生长于海拔 300～500 m 的石灰岩山地林中。

【附　　注】缺少影像资料。

[Scientific Name] *Kmeria septentrionalis*

[Morphological Characters] Trees，to 18 m tall，to 40 cm d.b.h. Bark gray. Twigs green，at first appressed pubescent. Leaves leathery，elliptic-oblong to obovate-oblong，8-15×3.5-6 cm，apex obtuse and slightly emarginate，base broadly cuneate，both surfaces glabrous or when young abaxially sparsely pilose at base，adaxially green and glossy， secondary veins 12-17 on each side of midvein, reticulate veins prominent on both surfaces why dry, petiole 2-3.5 cm, at first grayish pubescent, later glabrescent; stipule mark reaches petiole apex. Flower unisexual alien, male flower: tepals white and pale greenish; outer 3 tepals obovate, 2-3×ca. 2 cm; tepals of inner 2 whorls elliptic, slightly narrower and smaller; androecium white with pale yellow, obovoid; stamens 1.8-2.5 cm; filaments ca. 2 mm; connective exserted and forming a 2-3 mm tonguelike tip; anthers 1.5 cm, dehiscing laterally; female flower: outer 3 tepals obovate, 2.5-3×2-2.5 cm; tepals of inner whorls 8-10, linear-narrowly ovate, 2-2.5 cm×2-3 mm; gynoecium green, obovoid, ca. 10×8-9 mm; carpels 6-9; ovules 2 per carpel, shortly funiculate; styles short; stigmas cristate. Aggregate fruit subspherical and leathery, red when mature, subglobose; pericarp 3.5-4 cm, woody, dehiscing completely along dorsal sutures. Seeds 2 per carpel, those with testa red but those without testa black, bean-shaped to cordate, 10-12×7-9 mm, those without testa dark brown, apex truncate to slightly concave, narrowly long furrowed, with stalk in the middle, several irregular massive protuberances.

[Florescence and Fruit Period] Florescence May to June, fruit period October to November.

[Distribution] North of Guangxi Zhuang Autonomous Region (Luocheng, Huanjiang), Southeast of Guizhou Province (Libo).

[Habitat] Limestone mountain forest with an altitude of 300-500 m.

[Note] No image data.

68 落叶木莲/ Luoye Mulian

【学　　名】*Manglietia decidua*

【形态特征】落叶乔木，高约 15 m。芽及小枝无毛。叶革质，长圆状倒卵形、长圆状椭圆形或椭圆形，长 14～20 cm，先端钝或短尖，基部楔形，上面深绿色，无毛，下面粉绿色，初被白色丝状柔毛，后渐脱落，边缘微反卷；叶柄长 2.5～4.6（6）cm。花蕾具 1 佛焰苞状苞片；花梗长约 1 cm，初被柔毛。花黄色，花被片 15（16）片，外轮花被片长圆状椭圆形，长 7～7.4 cm，向内渐窄，内轮花被片披针形或线形，长 5.5～6 cm；雄蕊长 6～7 mm，花药长 4～5 mm；雌蕊群长约 1 cm，心皮 15～22 枚，每心皮 6～8 枚胚珠。聚合果卵圆形或近球形，长 4.7～7 cm，成熟时沿果轴从顶部至基部开裂，后反卷；蓇荚沿腹缝及几沿背缝全裂。种子红色。

【花 果 期】花期 4 月，果期 9—10 月。

【保护价值】数量稀少；珍贵；特有；受威胁严重；科学及文化价值大。

【致危因素】生境退化或丧失；物种内在因素。

[Scientific Name] *Manglietia decidua*

[Morphological Characters] Trees，deciduous，15 m tall. Buds and branchlets glabrous. Leaves leathery，oblong obovate，oblong elliptic or elliptic，14-20 cm long，apex blunt or acuminate，base cuneate，dark green adaxially，glabrous and pink green abaxially，white silky pilose at first，falling off gradually later，margin slightly retrorse；petiole 2.5-4.6（6）cm. Bud with 1 spathe bract；pedicel ca. 1 cm，initially pilose. Flowers yellow，perianth segments 15（16），outer perianth segments long round oval，7-7.4 cm long，gradually narrowed inward，inner perianth segments lanceolate or linear，5.5-6 cm；stamens 6-7 mm，anthers 4-5 mm；pistils group 1 cm，carpels 15-22，6-8 ovules per carpel. Fruit ovoid or subglobose，4.7-7 cm long，dehiscent from the top to the base along the fruit axis when it is mature，and then retrorse；the pods are completely dehiscent along the ventral seam and several dorsal seams. Seed red.

[Florescence and Fruit Period] Florescence April，fruit period September to October.

[Protection Value] Rare；precious；unique；seriously threatened；of great scientific and cultural significance.

[Threatening Factors] Habitat degradation or loss；internal factors of species.

摄像：朱鑫鑫

Photo：Zhu Xinxin

69 华盖木/ Huagaimu

【学　　名】 *Manglietiastrum sinicum*

【形态特征】 常绿大乔木，高达 40 m，胸径 1.2 m；树皮灰白色，细纵裂；干基部稍具板根；全株无毛。小枝深绿色，径 5～9 mm；老枝暗褐色。叶革质，狭倒卵形或狭倒卵状椭圆形，长 15～26（30）cm，宽 5～8（9.5）cm，先端圆，具长约 5 mm 的急尖，尖头钝而稍弯，基部渐狭楔形，下延，边缘稍背卷，上面深绿色，有光泽，下面淡绿色，中脉两面凸起，侧脉每边 13～16 条，网脉稀疏，干时两面均凸起；叶柄长 1.5～2 cm，无托叶痕，基部稍膨大。花单生枝顶，花蕾绿色，倒卵圆形或卵球形，佛焰苞状苞片紧接花被下；花被片 9，3 片 1 轮；外轮 3 片长圆状匙形，顶端钝，中轮及内轮 6 片，倒卵状匙形，较小；雄蕊约 65 枚，药室内向开裂，药隔伸出成长尖头；雌蕊群长卵球形，心皮 13～16 枚，每心皮具胚珠 3～5 枚。雌蕊群柄果时长约 1 cm。聚合果成熟时绿色，干时暗褐色，倒卵圆形、椭圆状卵圆形或倒卵圆形，长 5～8.5 cm，径 3.5～6.5 cm；蓇葖厚木质，狭长圆状椭圆体形或倒卵状椭圆体形，长 2.5～4 cm，径 1.5～2.5 cm，沿腹缝线全裂及顶端 2 浅裂，背面具粗皮孔；每心皮有种子 1～3 枚，种子横椭圆体形，两侧扁，宽 1～1.3 cm，高约 7 mm，腹孔凹入，中有凸点，背棱稍微凸。

【花 果 期】花期 4 月，果期 10—11 月。

【分　　布】产于云南（西畴法斗）。模式标本采自法斗。与大叶木莲、灯台树、伯乐树、刺楸、白克木等混交成林。

【生　　境】生长于海拔 1 300～1 500 m 的山沟常绿阔叶林中。

【保护价值】数量稀少；珍贵；特有；受威胁严重；科学及文化价值大。

【致危因素】物种内在因素；生境退化或丧失。

[Scientific Name] *Manglietiastrum sinicum*

[Morphological Characters] Trees，40 m tall，1.2 m d.b.h.，glabrous，bark grayish white，finely fissured，base slightly plate root，glabrous. Old twigs dark brown；young twigs deep green，5-9 mm in diam. Leaves leathery，narrowly obovate to narrowly obovate-elliptic，5-26（30）× 5-8（9.5）cm，apex rounded and with a ca. 5 mm acute tip，tip obtuse and slightly curved，base gradually narrowly cuneate and decurrent on petiole，margin slightly revolute，abaxially pale green，adaxially deep green and glossy，midvein prominent on both surfaces，secondary veins 13-16 on each side of midvein，reticulate veins sparse and prominent on both surfaces when dry；petiole 1.5-2 cm，without a stipular scar，base slightly enlarged. Flowers terminal，solitary；flower buds green，obovoid to ovoid. Spathaceous bracts just next to tepals. Tepals 9，in 3 whorls，apex blunt，outer whorl of 3 narrowly

ellipsoid，spoon-shaped，the middle wheel and the inner wheel are 6 pieces，obovate spoon shape，smaller；about 65 stamens，the medicine chamber is introrse and cracked，and the medicine septum is extended to the long acuminate；gynoecium long ovoid；carpels 13-16；ovules 3-5 per carpel. Fruiting gynophore ca. 1 cm. Fruit green when mature，dark brown when dry，obovoid to ellipsoid-ovoid，5-8.5×3.5-6.5 cm；follicles are thick woody，narrow，long，round or obovate ellipsoid，2.5-4×1.5-2.5 cm，with complete fissure along the abdominal suture and 2 shallow fissures at the top，with rough lenticels on the back；there are 1-3 seeds in each carpel，with transverse ellipsoid shape，flat on both sides，1-1.3 cm wide and about 7 mm high，with concave ventral holes，protruding points in the middle and slightly convex back edges.

[Florescence and Fruit Period] Florescence April，fruit October to November.

[Distribution] Yunnan Province（Xichou fadou）. The model specimens were collected from Fadou. Mixed with Manglietia grandiflora，Dengtai tree，Bole tree，acanthopanax，and Symingtonia populnea.

[Habitat] In the evergreen broad-leaved forest in the mountain valley with an altitude of 1,300-1,500 m.

[Protection Value] Rare，precious，unique，seriously threatened，of great scientific and cultural significance.

[Threatening Factors] Internal factors of species；habitat degradation or loss.

摄像：宋 鼎 朱鑫鑫 罗毅波

Photo：Song Ding Zhu Xinxin Luo Yibo

70 峨眉拟单性木兰/ Emei Nidanxingmulan

【学　　名】 *Parakmeria omeiensis*

【形态特征】 常绿乔木，高达 25 m，胸径 40 cm；树皮深灰色。叶革质，椭圆形、狭椭圆形或倒卵状椭圆形，长 8～12 cm，宽 2.5～4.5 cm，先端短渐尖而尖头钝，基部楔形或狭楔形，上面深绿色，有光泽，下面淡灰绿色，有腺点，侧脉每边 8～10 条，叶柄长 1.5～2 cm。雄花与两性花异株；雄花：花被片 12，外轮 3 片浅黄色较薄，长圆形，先端圆或钝圆，长 3～3.8 cm，宽 1～1.4 cm，内三轮较狭小，乳白色，肉质；倒卵状匙形，雄蕊约 30 枚，长 2～2.2 cm。花药长 1～1.2 cm，花丝长 2～4 mm，药隔顶端伸出成钝尖，药隔及花丝深红色，花托顶端短钝尖；两性花：花被片与雄花同，雄蕊 16～18 枚；雌蕊群椭圆体形，长约 1 cm，具雌蕊 8～12 枚。聚合果倒卵圆形，长 3～4 cm，种子倒卵圆形，径 6～8 mm，外种皮红褐色。

【花 果 期】 花期 5 月，果期 9 月。

【分　　布】 产于四川峨眉山。

【生　　境】 生长于海拔 1 200～1 300 m 的林中。

【保护价值】 数量稀少；珍贵；特有；受威胁严重；经济价值高；科学及文化价值大。

【致危因素】 物种内在因素；生境退化或丧失；种间影响。

[Scientific Name] *Parakmeria omeiensis*

[Morphological Characters] Trees, 25 m tall, 40 cm d.b.h.; bark dark grayish. Leaves leathery, elliptic, narrowly elliptic, or obovate-elliptic, 8-12×2.5-4.5 cm, apex shortly acuminate with an obtuse acumen, base cuneate to narrowly cuneate, green adaxially, glossy, gray-green abaxially, with glandular point, secondary veins 8-10 on each side of midvein, petiole 1.5-2 cm. Andro-dioecious, male flower: tepals 12; outer 3 tepals pale yellow, oblong, 3-3.8×1-1.4 cm, thinner, apex rounded or obtuse; tepals of inner 3 whorls milky white, obovate-spoon-shaped, fleshy, stamens ca. 30, 2-2.2 cm. Anthers 1-1.2 cm; filaments 2-4 mm, connective dark red, exserted and forming a mucro, torus apex mucronate. Bisexual flower: tepals same as those of male flowers; stamens 16-18; gynoecium ellipsoid, ca. 1 cm; carpels 8-12. Fruit obovoid, 3-4 cm. Seeds obovate, 6-8 mm in diam.; testa reddish brown.

[Florescence and Fruit Period] Florescence May, fruit period September.

[Distribution] Mount Emei, Sichuan Province.

[Habitat] In the forest with an altitude of 1,200-1,300 m.

[Protection Value] It is rare, precious, unique, seriously threatened, of high economic value and of great scientific and cultural significance.

[Threatening Factors] Intraspecific factors; habitat degradation or loss; interspecific influence.

摄像：李策宏

Photo: Li Cehong

71 藤枣/ Tengzao

【学　　名】*Eleutharrhena macrocarpa*

【形态特征】木质藤本，嫩枝有直线纹，被微柔毛，稍老即脱落。叶革质，卵形至阔卵形，长圆状卵形或长圆状椭圆形，长 9.5～22 cm，宽 4.5～13 cm，先端渐尖或近骤尖，基部圆或钝，有时阔楔形，两面无毛，上面光亮；侧脉 5～9 对，两面凸起，下面更高，网状脉稀疏，不很明显；叶柄长 2.5～8 cm。雄花序有花 1～3 朵，总梗长 6～10 mm，被微柔毛；雄花：外轮萼片微小，近卵形，长不及 1 mm，被微柔毛，中轮与外轮相似或稍长，内轮倒卵状楔形，长 1.5～1.7 mm，最内轮大，近圆形或阔卵状近圆形，长约 2.5 mm，无毛；花瓣 6 片，阔倒卵形，二侧边缘内卷，抱着花丝，无毛；雄蕊 6 枚，长约 1.5 mm。果序生无叶老枝上，总梗粗壮，长达 2 cm，其上有 3～6 个核果；核果椭圆形，黄色或红色，长 2.5～3 cm，宽 1.7～2.5 cm，心皮柄长达 1.5 cm；种子椭圆形，长 1.5～1.7 cm，宽约 1 cm。

【花 果 期】花期 5 月，果期 10 月。

【分　　布】产于云南南部和东南部。印度阿萨姆也有分布。

【生　　境】生长于海拔 840～1 500 m 的密林中，也见于疏林。

【致危因素】生境退化或丧失；直接采挖或砍伐。

[Scientific Name] *Eleutharrhena macrocarpa*

[Morphological Characters] Woody vines，young stems longitudinally striate，sparsely puberulent，later glabrous. Leaf blade leathery，ovate to broadly ovate，oblong-ovate，or oblong-elliptic，9.5-22×4.5-13 cm，apex acuminate or subcuspidate，base broadly rotund or obtuse，sometimes broadly cuneate，glabrous on both surfaces，adaxially glossy；lateral veins

5-9 pairs, prominent on both surfaces, especially abaxially, reticulation sparse and inconspicuous; petiole slender, 2.5-8 cm. Male inflorescences fasciculate, 1-3-flowered; peduncles 6-10 mm, puberulent. Male flowers: outer sepals minute, nearly ovate, less than 1 mm, puberulent, middle and outer rings similar or slightly longer, inner rings obovate cuneate, 1.5-1.7 mm long, the largest inner ring, nearly round or wide ovate nearly round, about 2.5 mm long, glabrous; petals 6, broadly obovate, with lateral margins inflexed and clasping opposite stamen, glabrous; stamens 6, ca. 1.5 mm. Drupes 3-6 on the leafless and old branches of the fruit sequence; drupes yellow or red, ellipsoidal, 2.5-3×1.7-2.5 cm, base with a gynophore up to 1.5 cm. Seeds ellipsoidal, 1.5-1.7×ca. 1 cm.

[Florescence and Fruit Period] Florescence May, fruit period October.

[Distribution] A native plant of South and Southeast Yunnan Province. Also found in Assam, India.

[Habitat] In dense forests with an altitude of 840-1,500 m and also in sparse forests.

[Threatening Factors] Degradation or loss of habitat; direct digging or cutting.

摄像：赵明旭　宋　鼎

Photo: Zhao Mingxu　Song Ding

72 莼菜/ Chuncai

【学　　名】*Brasenia schreberi*

【形态特征】多年生水生草本；根状茎具叶及匍匐枝，后者在节部生根，并生具叶枝条及其他匍匐枝。叶椭圆状矩圆形，长 3.5～6 cm，宽 5～10 cm，下面蓝绿色，两面无毛，从叶脉处皱缩；叶柄长 25～40 cm，和花梗均有柔毛。花直径 1～2 cm，暗紫色；花梗长 6～10 cm；萼片及花瓣条形，长 1～1.5 cm，先端圆钝；花药条形，约长 4 mm；心皮条形，具微柔毛。坚果矩圆卵形，有 3 个或更多成熟心皮；种子 1～2 枚，卵形。

【花 果 期】花期 6 月，果期 10—11 月。

【分　　布】江苏、浙江、江西、湖南、四川、云南。苏联、日本、印度、美国、加拿大、大洋洲东部及非洲西部均有分布。

【生　　境】生长在池塘、河湖或沼泽。

【保护价值】重要蔬菜，本种富胶质，嫩茎叶作蔬菜食用。

【致危因素】生境退化或丧失。

[Scientific Name] *Brasenia schreberi*

[Morphological Characters] Perennial aquatic herb; rhizome with leaves and stolons, the latter rooting at nodes and with leafy branches and other stolons. Leaves elliptical, oblong, 3.5-6×5-10 cm, blue-green abaxially, glabrous on both sides, shrunk from the vein; petioles 25-40 cm, pilose on both sides. Flowers 1-2 cm in diameter, dark purple; pedicels 6-10 cm in length; sepals and petals 1-1.5 cm in length, apex obtuse; anthers strip, about 4 mm in length; carpels strip, puberulent. Nuts oblong ovate, with 3 or more mature carpels; seeds 1-2, ovate.

[Florescence and Fruit Period] Florescence June, fruit period October to November.

[Distribution] Jiangsu Province, Zhejiang Province, Jiangxi Province, Hunan Province, Sichuan Province, and Yunnan Province. It is distributed in the Soviet Union, Japan, India, the United States, Canada, Eastern Oceania and Western Africa.

[Habitat] Pond, lake, or swamp.

[Protection Value] Important vegetables, this kind of rich gum, tender stem leaves for vegetable consumption.

[Threatening Factors] Habitat degradation or loss.

摄像：林秦文

Photo: Lin Qinwen

73 珙桐/ Gongtong

【学　　名】 *Davidia involucrata*

【形态特征】 落叶乔木，高 15～20 m，稀达 25 m；胸径约 1 m；树皮深灰色或深褐色，常裂成不规则的薄片而脱落。幼枝圆柱形，当年生枝紫绿色，无毛，多年生枝深褐色或深灰色；冬芽锥形，具 4～5 对卵形鳞片，常成覆瓦状排列。叶纸质，互生，无托叶，常密集于幼枝顶端，阔卵形或近圆形，常长 9～15 cm，宽 7～12 cm，顶端急尖或短急尖，具微弯曲的尖头，基部心脏形或深心脏形，边缘有三角形而尖端锐尖的粗锯齿，上面亮绿色，初被很稀疏的长柔毛，渐老时无毛，下面密被淡黄色或淡白色丝状粗毛，中脉和 8～9 对侧脉均在上面显著，在下面凸起；叶柄圆柱形，长 4～5 cm，稀达 7 cm，幼时被稀疏的短柔毛。两性花与雄花同株，由多数的雄花与 1 个雌花或两性花成近球形的头状花序，直径约 2 cm，着生于幼枝的顶端，两性花位于花序的顶端，雄花环绕于其周围，基部具纸质、矩圆状卵形或矩圆状倒卵形花瓣状的苞片 2～3 枚，长 7～15 cm，稀达 20 cm，宽 3～5 cm，稀达 10 cm，初淡绿色，继变为乳白色，后变为棕黄色而脱落。雄花无花萼及花瓣，有雄蕊 1～7 枚，长 6～8 mm，花丝纤细，无毛，花药椭圆形，紫色；雌花或两性花具下位子房，6～10 室，与花托合生，子房的顶端具退化的花被及短小的雄蕊，花柱粗壮，分成 6～10 枝，柱头向外平展，每室有 1 枚胚珠，常下垂。果实为长卵圆形核果，长 3～4 cm，直径 15～20 mm，紫绿色具黄色斑点，外果皮很薄，中果皮肉质，内果皮骨质具沟纹，种子 3～5 枚；果梗粗壮，圆柱形。

【花 果 期】花期 4 月，果期 10 月。

【分　　布】产于湖北西部、湖南西部、四川以及贵州和云南两省的北部。在四川西部的宝兴、天全、峨眉、马边、峨边等县极常见。

【生　　境】生长于海拔 1 500～2 200 m 的润湿的常绿阔叶落叶阔叶混交林中。

【保护价值】珙桐为古老、珍贵、稀有植物。盛花期头状花序下的 2 枚白色大苞片非常显著，极似展翅之群鸽栖于树上，故有“中国鸽子树”之称，是驰名世界的珍贵观赏树种。对研究古植物区系和系统发育均具有重要的科学价值。

【致危因素】生境退化或丧失。

[Scientific Name] *Davidia involucrata*

[Morphological Characters] Trees 15-20 m high，up to 25 m sparse，to 1 m d.b.h.；bark dark gray or dark brown，usually splitting into irregular flakes，falling off. Young branchlets cylindrical，the current year’s branches are purple green，glabrous，and the perennial branches are dark brown or dark gray；winter buds cone-shaped，with ovate scales 4-5 pairs，often arranged in tile like arrangement. Leaves papery，alternate，without stipules，often dense at the top of young branches，broad ovate or nearly round，9-15×7-12 cm wide，apex acuminate，slightly curved， heart-shaped or deep heart-shaped at the base，triangular and acuminate rough serrations at the margin，bright green adaxially，with sparse villous when young，glabrous when growing old，densely yellowish or pale white filamentous hirsute abaxially，midvein and 8-9 opposite veins are prominent above and convex below；petioles cylindrical，4-5 cm long，up to 7 cm sparse，sparsely pubescent when young. Bisexual flower growing with male flower，most male flowers and one female flower or bisexual flower form a nearly spherical flower head，2 cm in diam，with bisexual flower at inflorescence apex，surrounded by the male flower. At the base，there are basal 2-3 papery，oblong ovate or oblong obovate petal bracts，7-15 cm long，20 cm sparse，3-5 cm wide and 10 cm sparse，first pale green，then milky white，then brownish yellow and falling off. Male flowers without calyx and petals，with stamens 1-7，6-8 mm，filaments slender，glabrous，anther oval，purple；female or bisexual flowers with lower ovary，6-10 cells，with receptacle，the top of ovary with degenerate perianth and short stamen，style thick，divided into 6-10 branches，stigma outward flat，1 ovule in each cell，often pendulous. The fruit is a long oval drupe，3-4 cm long，15-20 mm in diam，purple green with yellow spots，very thin exocarp，fleshy mesocarp，sulcate endocarp，3-5 seeds；the fruit stem is thick and cylindrical.

[Florescence and Fruit Period] Florescence April，fruit period October.

[Distribution] West Hubei Province，West Hunan Province，Sichuan Province，North of Guizhou Province and Yunnan Province. It is very common in Baoxing，Tianquan，Emei，Mabian and Ebian counties in Western Sichuan Province.

[Habitat] In the mixed forest of moist evergreen broad-leaved and deciduous broad-leaved at an altitude of 1,500-2,200 m.

[Protection Value] Ancient，precious and rare. The two large white bracts under the flower heads are very significant during the blooming period. They are very similar to the group of pigeons that spread their wings and live on the trees，and are therefore called “Chinese pigeon trees”，and they are famous precious ornamental trees in the world. It is of great scientific value for the study of palaeoflora and phylogeny.

[Threatening Factors] Habitat degradation or loss.

摄像：林秦文

Photo：Lin Qinwen

74 光叶珙桐/ Guangye Gongtong

【学　　名】*Davidia involucrata* var.*vilmoriniana*

【形态特征】落叶乔木，高 15～20 m，稀达 25 m；胸径约 1 m；树皮深灰色或深褐色，常裂成不规则的薄片而脱落。幼枝圆柱形，当年生枝紫绿色，无毛，多年生枝深褐色或深灰色；冬芽锥形，具 4～5 对卵形鳞片，常成覆瓦状排列。叶纸质，互生，无托叶，常密集于幼枝顶端，阔卵形或近圆形，常长 9～15 cm，宽 7～12 cm，顶端急尖或短急尖，具微弯曲的尖头，基部心脏形或深心脏形，边缘有三角形而尖端锐尖的粗锯齿，上面亮绿色，叶背无毛或仅幼时具短柔毛，中脉和 8～9 对侧脉均在上面显著，在下面凸起；叶柄圆柱形，长 4～5 cm，稀达 7 cm，幼时被稀疏的短柔毛。两性花与雄花同株，由多数的雄花与 1 个雌花或两性花成近球形的头状花序，直径约 2 cm，着生于幼枝的顶端，两性花位于花序的顶端，雄花环绕于其周围，基部具纸质、矩圆状卵形或矩圆状倒卵形花瓣状的苞片 2～3 枚，长 7～15 cm，稀达 20 cm，宽 3～5 cm，稀达 10 cm，初淡绿色，继变为乳白色，后变为棕黄色而脱落。雄花无花萼及花瓣，有雄蕊 1～7 枚，长 6～8 mm，花丝纤细，无毛，花药椭圆形，紫色；雌花或两性花具下位子房，6～10 室，与花托合生，子房的顶端具退化的花被及短小的雄蕊，花柱粗壮，分成 6～10 枝，柱头向外平展，每室有 1 枚胚珠，常下垂。果实为长卵圆形核果，长 3～4 cm，直径 15～20 mm，紫绿色具黄色斑点，外果皮很薄，中果皮肉质，内果皮骨质具沟纹，种子 3～5 枚；果梗粗壮，圆柱形。

【花 果 期】花期 4 月，果期 10 月。

【分　　布】产于湖北西部、四川、贵州等地；常与珙桐混生。

【生　　境】生长于海拔 1 500～2 200 m 的润湿的常绿阔叶落叶阔叶混交林中。

【保护价值】是驰名世界的珍贵观赏树种，对研究古植物区系和系统发育均具有重要的科学价值。

【致危因素】生境退化或丧失。

[Scientific Name] *Davidia involucrata* var.*vilmoriniana*

[Morphological Characters] Trees 15-20 m high，up to 25 m sparse，to 1 m d.b.h.; bark dark gray or dark brown，usually splitting into irregular flakes，falling off. Young branchlets cylindrical，the current year's branches are purple green，glabrous，and the perennial branches are dark brown or dark gray；winter buds cone-shaped，with ovate scales 4-5 pairs，often arranged in tile like arrangement. Leaves papery，alternate，without stipules，often dense at the top of young branches，broad ovate or nearly round，9-15×7-12 cm wide，apex acuminate，slightly curved，heart-shaped or deep heart-shaped at the base，leaves abaxially glabrous or scarcely pubescent when young. triangular and acuminate rough serrations at the margin，bright green adaxially，with sparse villous when young，glabrous when growing old，densely yellowish or pale white filamentous hirsute abaxially，midvein and 8-9 opposite veins are prominent above and convex below；petioles cylindrical，4-5 cm long，up to 7 cm sparse，sparsely pubescent when young. Bisexual flower grow with male flower，most male flowers and one female flower or bisexual flower form a nearly spherical flower head，2 cm in diam，at young branchlets apex，with bisexual flower at inflorescence apex，surrounded by the male flower. Base papery，oblong ovate or oblong obovate petal bracts 2-3，7-15 cm long，20 cm sparse，3-5 cm wide and 10 cm sparse，first pale green，then milky white，then brownish yellow and falling off. Male flowers without calyx and petals，with stamens 1-7，6-8 mm，filaments slender，glabrous，anther oval，purple；female or bisexual flowers with lower ovary，6-10 cells，with receptacle，the top of ovary with degenerate perianth and short stamen，style thick，divided into 6-10 branches，stigma outward flat，1 ovule in each cell，often pendulous. The fruit is a long oval drupe，3-4 cm long，15-20 mm in diam，purple green with yellow spots，very thin exocarp，fleshy mesocarp，sulcate endocarp，3-5 seeds；the fruit stem is thick and cylindrical.

[Florescence and Fruit Period] Florescence April， fruit period October.

[Distribution] Western Hubei Province， Sichuan Province， Guizhou Province and other provinces； often mixed with Davidia involucrate.

[Habitat] In the mixed forest of moist evergreen broad-leaved and deciduous broad-leaved at an altitude of 1,500-2,200 m.

[Protection Value] A precious ornamental tree species well-known in the world；of important scientific value for the study of palaeoflora and phylogeny.

[Threatening Factors] Habitat degradation or loss.

摄像：徐晔春　李晓东

Photo：Xu Yechun　Li Xiaodong

75 云南蓝果树/ Yunnan Languoshu

【学　　名】 *Nyssa yunnanensis*

【形态特征】 大型乔木，高 25～30 m，胸径约 1 m；树皮深褐色，常现小纵裂；小枝粗壮，直径 5 mm，微呈棱角状，当年生枝密被黄绿色微绒毛，二年生以上枝被宿存的黄褐色微绒毛；皮孔显著，近圆形或椭圆形，淡白色或淡黄白色；冬芽锥形，鳞片镊合状排列，密被黄绿色绒毛。叶厚纸质，椭圆形或倒卵形，稀长椭圆形，长 15～22 cm，宽 8～12 cm，顶端钝尖，具短尖头，基部钝形或近圆形，稀楔形，边缘全缘或微呈浅波状，上面深绿色，干燥后橄榄色，下面除叶脉深黄色外其余部分淡绿色，干燥后灰绿色，密被黄绿色微绒毛，叶脉上更密，中脉在上面微下凹，在下面凸起，侧脉 14～18 对，与中脉成 40°展开，上部略向内弯曲；叶柄粗壮，长 2～3 cm，近圆柱形，上面微呈浅沟状，密被黄绿色微绒毛。花单性，异株，由叶腋或叶已脱落的叶痕腋芽生出；雄花多数成伞形花序，花梗长约 3 mm，被绒毛，总花梗粗壮，圆柱形，长 2～2.5 cm，密被黄绿

色微绒毛，单生于叶腋或叶已脱落的叶痕内侧；花托盘状；花萼有萼片5枚，卵形或三角状卵形，长约0.5 mm；外面被微绒毛，花萼下有小苞片4枚，卵形，密被绒毛；花瓣5枚，近长椭圆形，长2 mm，宽1 mm，外面被疏柔毛；雄蕊10枚，排列成2轮，长2～3 mm，生于花盘周围，花丝钻形，无毛，花药淡黄色，椭圆形；花盘肉质，微现裂痕。雌花未详。核果幼时绿色，干燥后紫褐色，长卵圆形或近椭圆形，长2 cm，宽1 cm，直径5 mm，被微绒毛，无果梗，通常4～5枚成头状果序，果实下边有矩圆形小苞片4枚；总果梗长2 cm，被黄绿色微绒毛，生于叶腋或叶已脱落的叶痕内侧；种子稍扁，外壳上有7条纵沟纹。

【花 果 期】花期3月下旬，果期9月。

【分　　布】产于云南南部。

【生　　境】生长于海拔500～1 100 m的山谷密林中。

【保护价值】我国热带北缘的特有种，对于区系研究有科学价值。木材坚重，为优良的建筑用材。

【致危因素】生境退化或丧失；物种内在因素。

[Scientific Name] *Nyssa yunnanensis*

[Morphological Characters] Trees dioecious, 25-30 m tall, ca. 1 m d.b.h. Bark dark brown, shallowly fissured; branchlets thick, 5 mm in diam, slightly angular, densely covered with yellow green microvilli in the current year, persistent yellow brown microvilli in the biennial and above; lenticels conspicuous, elliptic or obovate, pale white or yellowish white; winter buds cone-shaped, scales arranged in tweezers, densely covered with yellow green villi. Leaves papery, oval or obovate, rarely long oval, 15-22×8-12 cm, apex blunt with short mucro, base blunt or nearly rounded, rarely wedge-shaped, margin entire or slightly wary, dark green adaxially, turning olive when dry, light green on the lower part for the deep yellow vein, gray green after drying, dense yellow green microvilli, more dense on the vein, midvein slightly concave abaxially, lateral veins 14-18 pairs, at an angle of 40° with midvein, slightly curved inward at the upper part; petiole stout, 2-3 cm, nearly cylindrical, slightly gullied above, densely covered with yellow green microvilli. Flowers unisexual, alien, arising from axils of leaves or axillary buds of fallen leaves; male flowers mostly umbels, pedicels 3 mm, tomentose, total pedicels thick, cylindrical, 2-2.5 cm, densely covered with yellow green microvilli, which are solitary in the axils of the leaves or inside the fallen leaf scars; receptacle discoid; calyx with sepals 5, ovate or triangular ovate, about 0.5 mm long; outside covered with microvilli, under calyx with bracteoles 4, ovate, densely covered with villi; petals 5, nearly oblong, 2×1 mm, outside covered with sparsely pilose; stamens 10, arranged in 2 rounds, 2-3 mm long, around the disk, filaments subulate, glabrous, anthers yellowish, elliptic; disk fleshy and slightly cracked. Female flowers unknown. Drupe green when young, purple brown after drying, long oval or nearly elliptic, 2×1 cm, 5 mm in diameter, covered with microvilli, without fruit stalk, usually 4-5 in a capitate fruit sequence, with 4 oblong bracteoles under the fruit; the total fruit stalk 2 cm, covered with yellow green microvilli, growing in the axil of the leaf or inside the leaf mark where the leaf has fallen off; the seed is slightly flat, with 7 longitudinal grooves on the shell.

[Florescence and Fruit Period] Florescence late March, fruit period September.

[Distribution] South Yunnan Province.

[Habitat] In the dense forest of the valley with an altitude of 500-1,100 m.

[Protection Value] The endemic species of the northern margin of the tropics in China are of scientific value for floristic study. The wood is hard and heavy, which is a good building material.

[Threatening Factors] Habitat degradation or loss; internal factors of species.

摄像：慧肇祥　朱鑫鑫

Photo: Hui Zhaoxiang　Zhu Xinxin

76 合柱金莲木/ Hezhu Jinlianmu

【学　　名】*Sinia rhodoleuca*

【形态特征】直立小灌木，高约 1 m；茎常单生或近顶部分叉，暗紫色，光滑。叶薄纸质，狭披针形或狭椭圆形，长 7～15 cm，宽 1.5～3 cm，两端渐尖，边缘有密而不相等的腺状锯齿，两面光亮无毛，中脉两面隆起，侧脉多数，近平行，小脉明显；叶柄长 3～5 mm，腹面有槽。圆锥花序较狭，长 6～10 cm，花少数，具细长柄；萼片卵形或披针形，长 3～4 mm，浅绿色；花瓣椭圆形，长 4.5～6.5 mm，白色，微内拱；退化雄蕊宿存，白色，外轮的腺体状，基部连合成短管，中轮和内轮的长圆形，中轮的较大，顶端截平而有数小齿，内轮的略小，顶端微尖而具 3 齿裂；雄蕊长 2.5～3.5 mm，花丝短，花药箭头形，2 室；子房卵形，长约 2 mm，花柱圆柱形，柱头小，不明显。蒴果卵球形，长和宽约 5 mm，熟时 3 瓣裂；种子椭圆形，长约 1.7 mm，种皮暗红色，有多数小圆凹点。

【花 果 期】花期 4—5 月，果期 6—7 月。

【分　　布】产于广西北部（大苗山）、中部（大瑶山、象州）和广东西北部（怀集、封开、连山）。

【生　　境】生长于海拔 1 000 m 的山谷水旁密林中。

【保护价值】其形态特征与南美洲产的 *Wallacea* 属植物很接近，因此，它的存在对于进一步研究本科植物的区系、地理分布及其发生与发展等具有科学意义。

【致危因素】生境退化或丧失。

[Scientific Name] *Sinia rhodoleuca*

[Morphological Characters] Erect shrubs, ca. 1 m tall; stems often solitary or subterminal forked, dark purple, smooth. Leaves papery, narrowly lanceolate or narrowly elliptic, 7-15×1.5-3 cm, tapering at both ends, with dense but unequal glandular serrations at the edges, both sides glossy and glabrous, midvein raised at both surfaces, most are lateral veins, nearly parallel, with obvious veinlets; petioles 3-5 mm, grooved at the ventral side. Panicle narrow, 6-10 cm long, with flowers a few, stalks long and thin; sepals ovate or lanceolate, 3-4 mm, light green; petals oval, 4.5-6.5 mm, white, slightly arched inside; staminodes persistent, white, glandular in outer ring, base connected with synthetic short tube, oblong in middle ring and inner ring, large in middle ring, truncated at the top and small in inner ring, apex acuminate with 3-toothed cleft; stamens 2.5-3.5 mm, filaments short, anthers arrow shaped, 2-locular; ovary ovate, 2 mm, style cylindrical, stigma small, inconspicuous. Capsule ovoid, 5 mm in length and width, 3-valved at maturity; seeds elliptic, 1.7 mm, testa dark red, with many small pits.

[Florescence and Fruit Period] Florescence April to May, fruit period June to July.

[Distribution] Northwest (Mount Damiao), central (Mount Dayao, Xiangzhou) of Guangxi Zhuang Autonomous Region and Northwest of Guangdong Province (Huaiji, Fengkai, Lianshan).

[Habitat] In the dense forest beside the water in the valley, the altitude is 1,000 m.

[Protection Value] Its morphological character is very close to *Wallacea* in South America, so its existence is of scientific significance for further study of flora, geographical distribution, occurrence and development of undergraduate plants.

[Threatening Factors] Habitat degradation or loss.

摄像：张金龙　徐晔春

Photo: Zhang Jinlong　Xu Yechun

77 独叶草/ Duyecao

【学　　名】*Kingdonia uniflora*

【形态特征】多年生小草本，无毛。根状茎细长，自顶端芽中生出 1 叶和 1 条花葶；芽鳞约 3 个，膜质，卵形，长 4～7 mm。叶基生，有长柄，叶片心状圆形，宽 3.5～7 cm，五全裂，中、侧全裂片三浅裂，最下面的全裂片不等二深裂，顶部边缘有小牙齿，背面粉绿色，叶柄长 5～11 cm。花葶高 7～12 cm。花直径约 8 mm；萼片（4）5～6（7），淡绿色，卵形，长 5～7.5 mm，顶端渐尖；退化雄蕊长 1.6～2.1 mm；雄蕊长 2～3 mm，花药长约 0.3 mm；心皮长约 1.4 mm，花柱与子房近等长。瘦果扁，狭倒披针形，长 1～1.3 cm，宽约 2.2 mm，宿存花柱长 3.5～4 mm，向下反曲，种子狭椭圆球形，长约 3 mm。

【花 果 期】5—6 月开花。

【分　　布】分布于云南西北部（德钦）、四川西部、甘肃南部（舟曲）、陕西南部（太白山）。

【生　　境】生长于海拔 2 750～3 900 m 的山地冷杉林下或杜鹃灌丛下。

【保护价值】数量稀少；珍贵；特有；受威胁严重；科学及文化价值大。

【致危因素】个体数量少，生境退化或丧失。

[Scientific Name] *Kingdonia uniflora*

[Morphological Characters] Small perennial herbs, glabrous. Rhizome slender, with 1 leaf and 1 scape from the top bud; bud scales 3, membranous, ovate, 4-7 mm. Leaves basal, with long stalks, round heart-shaped leaves, 3.5-7 cm, full splits 5, three shallow splits of middle and side full lobes, basal full lobes unequal to two deep splits, with small teeth at apex margin, flour green abaxially, petiole 5-11 cm. Scape 7-12 cm in height. Flower 8 mm in diam; sepals (4) 5-6 (7), light green, ovate, 5-7.5 mm in length, apex acuminate; staminodes 1.6-2.1 mm; stamens 2-3 mm, anthers 0.3 mm; carpels 1.4 mm, style nearly equal to ovary. Achenes flat, narrowly oblanceolate, 1-1.3 cm×2.2 mm, persistent styles 3.5-4 mm, recurved downward, seeds narrowly ellipsoid, 3 mm long.

[Florescence and Fruit Period] Fruit period May to June.

[Distribution] Northwest Yunnan Province (Deqin), West Sichuan Province, South Gansu Province (Zhouqu), south Shaanxi Province (Mount Taibai).

[Habitat] Between 2,750-3,900 m above sea level, under fir forest or Rhododendron bush in mountainous area.

[Protection Value] Rare, precious, unique, seriously threatened, of great scientific and cultural significance.

[Threatening Factors] Small number of individuals, habitat degradation or loss.

摄像：朱仁斌

Photo: Zhu Renbin

参考文献

[1] 于永福. 中国野生植物保护工作的里程碑——《国家重点保护野生植物名录(第一批)》出台[J]. 植物杂志，1999（5）：3.

[2] 国家重点保护野生植物名录（第一批）[J]. 植物杂志，1999（5）：4-11.

[3] 国家林业局野生动植物保护和自然保护区管理司，中国科学院植物研究所. 中国珍稀濒危植物图鉴[M]. 北京：中国林业出版社，2013.

[4] 中国科学院中国植物志编辑委员会. 中国植物志[M]. 北京：科学出版社，2004.

致　谢

本书在编写过程中得到热心人士的支持与帮助，在此一并表示感谢！

特别感谢张志翔（北京林业大学）、尚策（北京林业大学）对本书的编写提出的宝贵建议。感谢林秦文（中国科学院北京植物所）、汪远（北京世界园艺博览会植物馆）、毛星星（四川大学）、王启平（安徽古井原国家级自然保护区管委会）等为本书提供的精美照片。